Philipp Mania

Precision Synthesis of Heterogeneous Epoxidation Catalysts based on Ti

Philipp Mania

Precision Synthesis of Heterogeneous Epoxidation Catalysts based on Ti

Investigation of molecular surface processes

Südwestdeutscher Verlag für Hochschulschriften

Impressum / Imprint
Bibliografische Information der Deutschen Nationalbibliothek: Die Deutsche Nationalbibliothek verzeichnet diese Publikation in der Deutschen Nationalbibliografie; detaillierte bibliografische Daten sind im Internet über http://dnb.d-nb.de abrufbar.
Alle in diesem Buch genannten Marken und Produktnamen unterliegen warenzeichen-, marken- oder patentrechtlichem Schutz bzw. sind Warenzeichen oder eingetragene Warenzeichen der jeweiligen Inhaber. Die Wiedergabe von Marken, Produktnamen, Gebrauchsnamen, Handelsnamen, Warenbezeichnungen u.s.w. in diesem Werk berechtigt auch ohne besondere Kennzeichnung nicht zu der Annahme, dass solche Namen im Sinne der Warenzeichen- und Markenschutzgesetzgebung als frei zu betrachten wären und daher von jedermann benutzt werden dürften.

Bibliographic information published by the Deutsche Nationalbibliothek: The Deutsche Nationalbibliothek lists this publication in the Deutsche Nationalbibliografie; detailed bibliographic data are available in the Internet at http://dnb.d-nb.de.
Any brand names and product names mentioned in this book are subject to trademark, brand or patent protection and are trademarks or registered trademarks of their respective holders. The use of brand names, product names, common names, trade names, product descriptions etc. even without a particular marking in this works is in no way to be construed to mean that such names may be regarded as unrestricted in respect of trademark and brand protection legislation and could thus be used by anyone.

Coverbild / Cover image: www.ingimage.com

Verlag / Publisher:
Südwestdeutscher Verlag für Hochschulschriften
ist ein Imprint der / is a trademark of
OmniScriptum GmbH & Co. KG
Heinrich-Böcking-Str. 6-8, 66121 Saarbrücken, Deutschland / Germany
Email: info@svh-verlag.de

Herstellung: siehe letzte Seite /
Printed at: see last page
ISBN: 978-3-8381-3802-2

Zugl. / Approved by: Zürich, ETH, Diss., 2013

Copyright © 2014 OmniScriptum GmbH & Co. KG
Alle Rechte vorbehalten. / All rights reserved. Saarbrücken 2014

"Oxygen gets you high. In a catastrophic [flight] emergency, you're taking giant panicked breaths. Suddenly you become euphoric, docile. You accept your fate. It's all right here."

Tyler Durden
Fight Club (1999)

Acknowledgements

I am very grateful to Prof. Dr. Ive Hermans for supervising my doctoral studies at the Institute for Chemical and Bioengineering (ICB). I highly appreciated Professor Hermans' research advice, motivating me for the study of complex oxidation reactions. His mentoring made sure that the projects went in promising directions. His profound knowledge of chemistry and reaction engineering was inspiring for my own scientific endeavor.

In addition to the group members, I would like to thank all the student who chose me for one of their projects.

I would like to thank Sebastian for proof-reading and for all his numerous visits in Zurich despite the long distance.

A lot of thanks need also be addressed to all my other friends in Zurich who made this very long stay unforgettable and I will hopefully stay in contact with.

Finally I would like to thank my parents and my sister for their never ending support and believe in me. Likewise, I want to thank Christin for always following my steps in life and never cutting my ties.

Articles

- „Understanding selective oxidations"
  U. Neuenschwander, N. Turrà, C. Aellig, P. Mania, I. Hermans
  CHIMIA **2010**, *64*, 225-230.
  (chapter 1)

- „Thermal Restructuring of Silica Grafted $TiCl_x$ Species and Consequences for Epoxidation Catalysis"
  Philipp Mania, René Verel, Forian Jenny, Ceri Hammond, Ive Hermans
  accepted in Chem. Eur. J., DOI: 10.1002/chem.200
  (chapter 2)

- „ Thermal Restructuring of Silica-Grafted $-CrO_2Cl$ and $-VOCl_2$ Species"
  Philipp Mania, Sabrina Conrad, René Verel, Ceri Hammond, Ive Hermans
  accepted in Dalton, DOI: 10.1039/c0xx00000x
  (chapter 3)

Conferences

- P. Mania, I. Hermans
  Session: Catalyst Science and Engineering
  "Tuning chemical vapor deposited d^0 Lewis acid sites on silica"
  SCS Fall Meeting, Zürich, Sep. 2012

- P. Mania, I. Hermans
  Prepr. Pap.-Am. Chem. Soc., Div. Catl. Chem. **2012**
  „Thermal Alternation of d^0 Transition Metal Lewis Sites Deposited onto Silica"
  244th ACS Fall Meeting, Philadelphia, PA, Aug. 2012

- P. Mania, I. Hermans
  Prepr. Pap.-Am. Chem. Soc., Div. Pet. Chem. **2012**, *57(1)*, 248
  „Thermal Restructuring of Silica-Grafted Ti^{IV} Lewis Acid Sites"
  243rd ACS Spring Meeting, San Diego, CA, Mar. 2012

- P. Mania, I. Hermans
  "Thermal Restructuring of Silica-Grafted Ti^{IV} Lewis Acid Sites"
  122th International Summer Course at BASF, Ludwigshafen, Aug. 2011

Table of Contents

Abstract 1

Zusammenfassung 5

I Introduction 9

1 Introduction to Epoxidations 10
1.1 Industrial perspectives 11
1.2 The fundamentals of Epoxidation Chemistry 14
1.3 Choice of the Oxidation Agent 17
1.4 Homogeneous Catalytic Epoxidation 21
1.5 Conclusions 41
1.6 Heterogeneous Catalytic Epoxidation 43
1.7 Conclusions 64

II Synthesis of Heterogeneous d^0 Lewis Acid Catalysts 67

2 Grafting TiCl$_4$ onto Silica 68
2.1 Introduction 69
2.2 Results and Discussion... 71
2.3 Conclusions 98

3 Grafting VOCl$_3$ and CrO$_2$Cl$_2$ onto silica 100
3.1 Introduction 101
3.2 Results and Discussion... 102
3.3 Conclusions 128

4 Synthesis of a dimeric Ti site catayst................................... 130
4.1 Introduction 131
4.2 Results and Discussion... 132
4.3 Conclusions 141

III Kinetic Investigations 143

5 Kinetic Experiments with *tert*-Butyl Hydroperoxide 144
5.1 Introduction 145
5.2 Catalytic Epoxidation with *tert*-Butyl Hydroperoxide 147
5.3 Conclusions 154

6 Kinetic Experiments with hydrogen peroxide 156
6.1 Introduction 157
6.2 Catalytic Epoxidation with Hydrogen Peroxide 160
6.3 Decomposition of Hydrogen Peroxide 168
6.4 Conclusions 179

IV Outlook 182

7 Outlook 183

V Appendix 189

8 Experimental 190
9 References 204
10 List of Abbreviations and Acronyms 225

Abstract

In this PhD work, a catalyst was prepared by chemical vapor deposition (CVD) of $TiCl_4$ onto Aerosil200 serving as an amorphous SiO_2 source, similar to the synthesis of Shell's SMPO process, and presumably the titanium site in TS1. The synthesis procedure consists of three elementary steps: 1. Thermal pretreatment; 2. Reaction and desorption phase; 3. Thermal Posttreatment. It was found that desorption time and the amount of $TiCl_4$ deposited had no influence on the final Ti surface concentration as long as the $TiCl_4$/SiOH ratio was above 1. All steps were conducted under high vacuum in order to avoid contact with moisture. For the same reason, all materials were stored in a glove box (H_2O < 1 ppm) due to their fast hydrolysis. In the first step the plain silica was taken and heated to 700°C to obtain an amorphous silica with solely isolated surface silanol groups ($\equiv$Si-OH). In a second step, the isolated OH-groups are brought into contact with $TiCl_4$ in order to yield only monopodal surface species ($\equiv$SiO-$TiCl_3$, see fig. 1). After desorbing excess $TiCl_4$, the silica sample with the monopodal Ti species was heated to 450°C which resulted in a decrease of the Ti content of almost 50%. Remarkably, the effect of the thermal posttreatment is barely mentioned in the literature, especially devoid of moisture, and was therefore investigated in more detail.

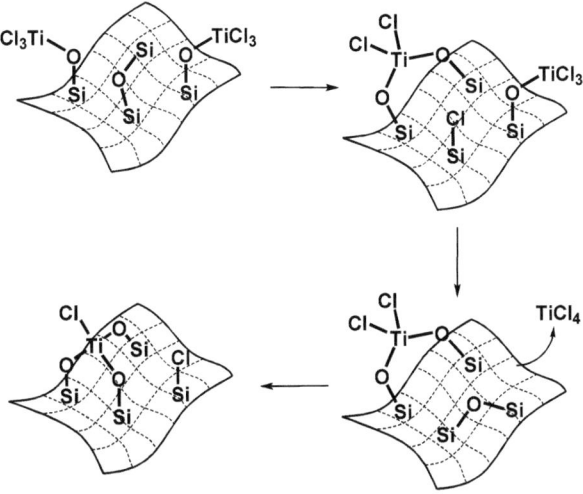

Figure 1: Hypothesis of surface restructuring of titanium surface species

Because of the decreases in Ti loading and Cl/Ti ratio a surface elimination of $TiCl_4$ was hypothesized. The ^{35}Cl NMR of three samples T50, T250 and T450, heated at different at post-treatments of 50, 250 and 450°C respectively, gave evidence for two different compounds, characterized by a narrow and a broad signal. The reason for the remarkably narrow signal must be an averaging due to a molecular rotation of the $TiCl_3$-unit around the Ti-O-Si-bond at room temperature which one could expect because the molecule is linked to a solvent-free surface. Quantum chemistry calculations for such a rotating surface species predicted a quadropole coupling constant of 5.7 MHz which is in very good agreement with the experimentally observed one of 4 MHz. Additionally the computational predictions of the UVVIS spectra for the three hypothesized species are very similar to the experimental UVVIS spectra of samples T50, T250 and T450. XPS, EXAFS and RAMAN spectroscopy supported our surface restructuring hypothesis.

In order to evaluate the influence of the posttreatment on the activity in olefin epoxidation, three catalyst were prepared and tested. Two catalyst were synthesized according to the procedure described above with posttreatment temperatures of 250 (T250) and 450°C (T450), and one stat-of-the-art catalyst[1]. The epoxidation was done at 80°C in pure octene and with 100 mM *tert*-Butyl hydroperoxide in continuous-flow reactor. The T450 proved to be the most active and stable material under those conditions and its performance was therefore evaluated for different olefins. Moreover, an additional catalyst was prepared, as analogous to T450 but in a subsequent synthesis step titanium tetra-*iso*-propoxide was grafted onto its surface. After this impregnation step, the titanium content doubled, and no more chlorine could be found on the catalyst surface as evidenced by ICP and EXAFS. Those observations seem to suggest the formation of a Ti-O-Ti moiety and the hypothesis could be tested if a dimer site is more efficient in olefin epoxidation. The activation barriers for different olefins were evaluated over the mono- and binuclear catalysts but no significant difference could be found. However, the T450 showed a higher activity per Ti site than the dimeric Ti species.

The T450 was also tested for olefin epoxidations with hydrogen peroxide and compared to titanium silicate 1 (TS1) which is used industrially (HPPO process of BASF-Dow). The activity of T450 in aqueous H_2O_2 was drastically decreased due to the following effects: the presence of water and the decomposition of H_2O_2 on acidic silanol groups formed under the reaction conditions. In an experiment with increasing water concentration the activation barrier for olefin epoxidation did increase. This is pointing towards a changing activation efficiency, probably due to coordinated water which supports the hypothesis that the hydrophobic pores of the TS1 create an optimal reaction environment for the epoxidation. When H_2O_2 was added drop-wise to a batch epoxidation, the conversion could be increased which is pointing towards undesired decomposition of the oxidant. Investigations of the decomposition of H_2O_2 showed that the silanol groups on the silica surface are the main contributors. Simple silylation resulted in a total deactivation for H_2O_2 decomposition of the silica surface. It could also be shown that the decomposition only plays an important role for temperatures above 60°C, making the T450 a potential catalyst for olefin epoxidation with hydrogen peroxide under moderate conditions.

In a side-project, the behavior of $VOCl_3$ and CrO_2Cl_2 grafted on silica were investigated during the same posttreatment as for $TiCl_4$. Only CrO_2Cl_2 did form a similar bipodal species as in the case for $TiCl_4$. Due to its low Tamman temperature, the vanadium integrated into the loose silica network building isolated, tetrahedrally coordinated sites.

Zusammenfassung

In dieser Doktorarbeit wurden Katalysatoren hergestellt mithilfe des Verfahren „Chemical Vapor Deposition" (CVD) von $TiCl_4$ auf Aerosil200, einem armorphen Siliziumoxid. Die Synthese war ähnlich Shells SMPO Prozeß und sollte Titanspezies ähnlich denen im TS1 hervorbringen. Die Herstellung bestand aus drei grundsätzlichen Schritten: 1. Thermische Vorbehandlung; 2. Reaktions- und Desorptionsphase; 3. Thermische Nachbehandlung. Die Desorptionszeit und die Menge an transferiertem $TiCl_4$ aus Schritt 2 hatten keinen Einfluss auf die finale Titan-Oberflächenkonzentration, solange das Verhältnis von transferiertem $TiCl_4$ und SiOH-Gruppen größer als 1 war. Alle Schritte wurden in hohem Vakuum ausgeführt, um den Kontakt jeglicher Art mit Wasser (z.B. Luftfeuchtigkeit) zu vermeiden. Aus dem gleichen Grund wurden all hergestellten Materialien in einer Glove-Box box (H_2O < 1 ppm) aufbewahrt, da die hergestellten Oberflächenspezien sonst schnell hydrolysierten.

Im ersten Schritt wurde das Siliziumoxid auf 700°C in Vakuum erhitzt, um einen armorphen Träger mit isolierten Silanolgruppen ($\equiv$Si-OH) zu erhalten. Im zweiten Schritt wurden diese mit $TiCl_4$ in Kontkat gebracht, um ausschließlich monopodale Oberflächenspezies species ($\equiv$SiO-$TiCl_3$, see fig. 1) herzustellen. Nach Desorption des überschüßigen $TiCl_4$ wurde die Probe mit der monopodalen Titanspezies auf 450°C erhitzt, welches in einer Erniedrigung des Titangehlats zu fast 50% führte. Überraschenderweise wurde dieser Nachbehandlungseffekt in der Literatur bisher kaum erwähnt, besonders unter Ausschuß von Feuchtigkeit, und wurde daher genauer untersucht.

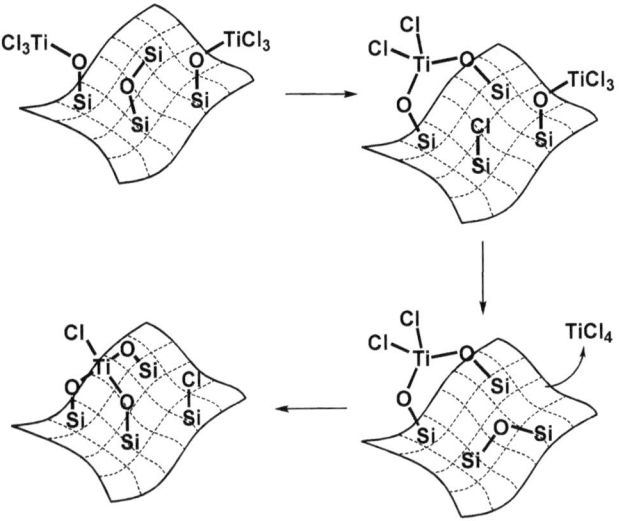

Figur 2: hypothetische Oberflächen-Umstrukturierung der Titan-Oberflächenspezies

Der Verlust an Titangehalt und das niedrigere Cl/Ti-Verhältnis wurde durch eine Oberflächeneliminierung erklärt. Die ^{35}Cl NMR-Spektren der Proben T50, T250 und T450, nachbehandelt bei Temperaturen von jeweils 50, 250 und 450°C, zeigten zwei verschiedene Signale, einem schmalen und einem breiten. Der Grund für das überaus schmale Signal muss durch ein Ausmitteln aufgrund einer molekularen Rotation der TiCl$_3$-Einheit um die Ti-O-Si-Achse bei Raumtemperatur, welche man erwarten würde, da das Molekül monopodal auf einer lösungsmittelfreien Oberfläche befestigt ist. Quantenchemische Berechnungen zu diesen rotierenden Oberflächenspezien ergaben eine Quadrupolkopplungskonstante (QKK) von 5.7 MHz, welche mit der experimentellen QKK von 4 MHz sehr gut übereinstimmte. Zusätzlich deckten sich die UV-VIS-Spektren der drei berechneten Titan-Oberflächenspezien (mono-, bi- and tripodal, siehe Fig. 1) sehr gut mit den experimentell bestimmten der drei Proben (T50, T250 und T450). XPS, EXAFS und RAMAN-Spektroskopie unterstützen die Hypothese der Oberflächenumstrukturierung ebenfalls.

Um den Einfluss der Nachbehandlungstemperatur auf die Aktivität in Olefin-Epoxidierungen zu untersuchen, wurden drei verschiedene Katalysatoren hsynthetisiert und getestet. Zwei Katalysatoren wurden hergestellt nach dem oben genannten CVD-Prozeß mit Nachbehandlungstemperaturen von 250 (T250) und 450°C (T450), während der dritte Katalysator hergestellt wurde nach einem Standardverfahren aus der Literatur[1]. Die Epoxidierung wurde bei 80°C in reinem Cyclookten und 100mM *tert*-Butyl hydroperoxid in einem Continuous-Flow-Reaktor durchgeführt. T450 war das aktivste und stabilste Material unter diesen Bedingungen, weshalb er für weitere Untersuchungen mit anderen Olefinen getestet wurde. Außerdem wurde ein weiterer Katalysator synthetisiert, basierend auf dem T450. Dazu wurde in einem zusätzlichen Schritt auf den T450-Katalysator selbst Titan-iso-propoxid abgeschieden. Nach diesem Impregnationsschritt verdoppelte sich der Titangehalt und jegliches Chlor wurde nachweislich durch EXAFS und ICP entfernt. Diese Beobachtungen bestärkten die erhoffte Bildung einer Ti-O-Ti-Einheit und dadurch konnte die Hypothese getestet werden, ob eine dinukleare Titaneinheit Olefine effizienter epoxidiert. Die Aktivierungsenergien der getesteten Olefine mit der mono- und binuklearen Titaneinheit zeigten keinen signifikanten Unteschied, wobei T450 jedoch aktiver pro Titanatom war als der binukleare Katalysator.

T450 wurde auch in der Olefin-Epoxidierung mit Wasserstoffperoxid getestet und mit dem industriellen Pendant Titansilcat-1 (TS1) verglichen, welches Einsatz im Dow/BASF HPPO-Prozeß Anwendung findet. Die Aktivität des T450 in wässrigem Wasserstoffperoxid war drastisch niedriger aufgrund der folgenden Effekte: Das wässrige Milieu und die Zersetzung von Wasserstoffperoxid an sauren Silanolgruppen, welche *in-situ* gebildet wurden. Die Aktivierungenergie erhöhte sich in Experimenten mit erhöhten Wassergehalt. Diese Beobachtung zeigt, dass die Anwesenheit von Wasssrer die Aktivierung erschwert und unterstützt die Hypothese, wonach die hydrophoben Poren des TS1 eine optimale Reaktionsumgebung zur Epoxidierung mit wässrigem H_2O_2 bilden. Batch-Experimente mit einer sequentiellen Zugabe des H_2O_2 und mit sylyliertem Siliziumoxid zeigten, dass die Silanolgruppen auf der Siliziumoberfläche den größten Anteil an der Zersetzung des Wasserstoffperoxids ausmachen. Sylierung der Silanolgruppen resultierte in der kompletten Deaktivierung der Silanolgruppen in der H_2O_2-Zersetzung. Außerdem wurde gezigt, dass die Zersetzung nur eine Rolle bei Temperaturen über 60°C spielt. Das bedeutet, dass T450 ein hoffnungsvoller Kandidat zur Epoxidierung mit Wasserstoffperoxid unter milden Bedingungen ist.

In einem Nebenprojekt wurde das Verhalten von Materialien, die mit $VOCl_3$ und CrO_2Cl_2 imprägniert wurden, während der gleichen Nachbehandlung wie beim T450 untersucht. Nur CrO_2Cl_2 zeigt ein ähnliches Verhalten und bildete bipodale Oberflächenspezien. Aufgrund der niedrigen Tammann-Temperatur lagerte sich das Vandium in das lose Siliziumoxid-Netzwerk ein und bildete isolierte, tetrahedrische Vanadiumoxide.

Part II

Introduction

1 Introduction to Epoxidations

Introducing oxygen functionalities into hydrocarbons is a very important reaction in the chemical industry and has vast influence in many economical fields. One of the many areas in these transformations is the epoxidation of olefins. The resulting epoxides are intermediates along the pathway of chemical synthesis leading from the raw to the final product and thereby adding economic value. For instance, ethene oxide is produced on an 18 Mt scale per year and propene oxide on an 8 Mt scale per year[2]. The main obstacles in this field are the efficient use of the oxidant. Normally hydrogen peroxide would be the first and ideal choice, but it suffers from decomposition and detailed insight into the reaction mechanism, amongst other problems such as over-oxidation and parallel side reactions. In this chapter, a short overview of different oxidation reactions is given in order to highlight the achievements made in this field of study.

1.1 Industrial Perspectives

Chemical industry is at the heart of modern energy management and production of daily goods. Since the limitation of the petrochemical resources is shifting more and more into the focus of industrial production, new key points on which chemists and engineers are concentrating are renewable feedstocks; reusing side or by products; decreasing waste and matching the increasing demand for products and energy. These new challenges of our industrialized civilization need to be integrated into our existing production chain. New more sustainable solutions need to be developed in which chemistry represents one of the main strategic components[1]. Even though economic and ecological incentives are often contrary to each other it needs to be kept in mind that process optimizations are usually driven by less energy consumption and/or smaller production of waste, which are beneficial in each way. Sometimes a small improvement can eventually spark off a whole series of positive effects. *E.g.*, lowering the energy barrier of one desired reaction pathway due to the use of a superior catalyst does not only increase the product output but reduces the amount of waste and simplifies the post-reaction separation.

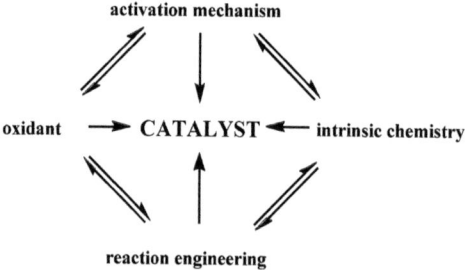

Figure 1: Different parameters determining the performance of an oxidation process.

From a sustainability point of view, selective oxidations are best performed in the presence of a catalyst to mediate the activity and selectivity. This prevents the formation of waste, and therefore reduces the downstream separation costs making heterogeneous catalysis an often used concept in industry. Some examples of mass scale productions are the industrial synthesis of terephthalic acid, formaldehyde and ethene oxide. In addition, fine chemicals are prepared catalytically in a much smaller scale, *e.g.*, for fragrance and parfume production[2]. Due to the fact that petrochemical resources are limited, a lot of effort has been put in the development of, *e.g.*, heatrecovery of exothermic reactions or replacement of environmentally risky processes by integrating new chemical solutions[2].

Selective oxidations are controlled by different parameters (Figure 1). One of them is the intrinsic chemistry of the specific oxidation reaction. The knowledge of the different reaction steps and its interaction with one another determines the complete reaction mechanism and its chemical outcome. Due to the combination of this knowledge the catalytic system can be improved, thus influencing the choice of the oxidant, activation energy and the reaction engineering.

Another cornerstone of Fig. 1 is the activation mechanism which is intrinsically connected to the engineering of the reaction because the activation requires a source of energy such as light, heat or electricity. In order to supply this activation energy a unique, industrial site-dependent and integrated solution needs to be found depending on economic demands and the technology provided in a plant.

In the center of the whole process is the catalyst, converting the reactants to the economically more valuable products. Its structure and stability determine its economic and ecological value whereby heterogeneous catalysts have the advantage of a convenient post-reaction separation over its homogeneous counterparts. In order to meet all demands the above mentioned properties are of foremost importance[3].

1.2 The Fundamentals of Epoxidation Chemistry

Oxidations can involve hemolytic or heterolytic bond cleavage or both depending on the oxidant and reactants used.[4a] A homolytic bond cleavage is often triggered by radicals (*e.g.*, O_2) and produces radicals resulting in a catalytic cycle which can only be terminated by a reaction of two radicals or an one-electron transfer. A heterolytic bond cleavage is often mediated by the use of a (transition) metal, *e.g.*, Ti^{IV}, V^V or Cr^{VI}, and usually passes through a two-electron transfer as in conventional RedOx-Chemistry.

One interesting example of a heterolytic reaction is the epoxidation of C=C double bonds. The formed three-membered-heterocycle is very strained (27 kcal/mol) and can be used as a starting point for the synthesis of a broad range of molecules[4b]. The epoxidation can be conducted via a nucleophilic or electrophilic attack of the oxidant. The first of which is the more extensively studied reaction (*e.g.*, the Sharpless oxidation[4c]). Depending on the catalyst and oxidant applied, different mechanisms are proposed. They mostly have in common to conduct the reaction from the double bond to an electrophilic O-donor (*e.g.*, ozone, hypochlorite, periodates or peroxides) via a two-electron transfer:

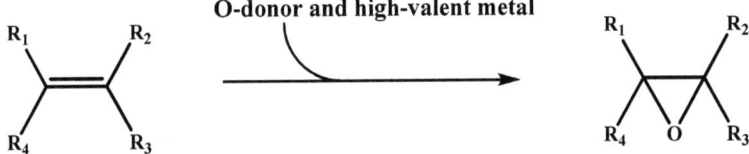

Figure 2: Epoxidation via nucleophilic attack on the olefin.

The reaction is usually catalyzed by early (M = Mo, W, Ti, V, Cr, Re) or late transition metals (M = Fe, Mn, Ru, Ni, Pd, Pt, Co, Cu). The former group catalyzes epoxidation via a metal-peroxo-species (Figure 3) whereas the latter group forms reactive metal-oxo-species [O=M] which can transfer the electrophilic oxygen onto the double bond. Therefore the reaction works best for electron-rich (higher substituted) olefins[4d] even though steric hindrance can be an issue. Another problem can be over-oxidation mostly due to the very active O-donors or the metal oxo/peroxo species.

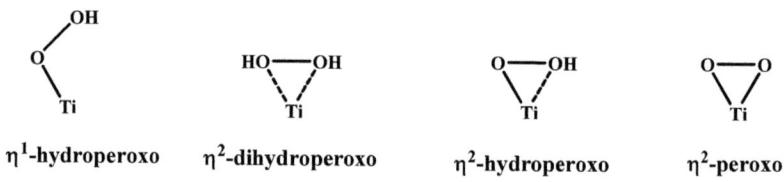

η^1-hydroperoxo η^2-dihydroperoxo η^2-hydroperoxo η^2-peroxo

Figure 3: Different coordinations of hydrogen peroxide to titanium.

Another mechanism (Weitz-Scheffer epoxidation[4e]) is also occurring through a two-electron transfer but the employed O-donor attacks the olefin nucleophilically (Fig. 3), while the group attached to the nucleophilic oxygen (LG) is leaving the reactant in a concerted mechanism[4e]. The reaction is not catalyzed by a metal *per se* but introducing chiral metal compounds for asymmetric epoxidation has been verified[4f].

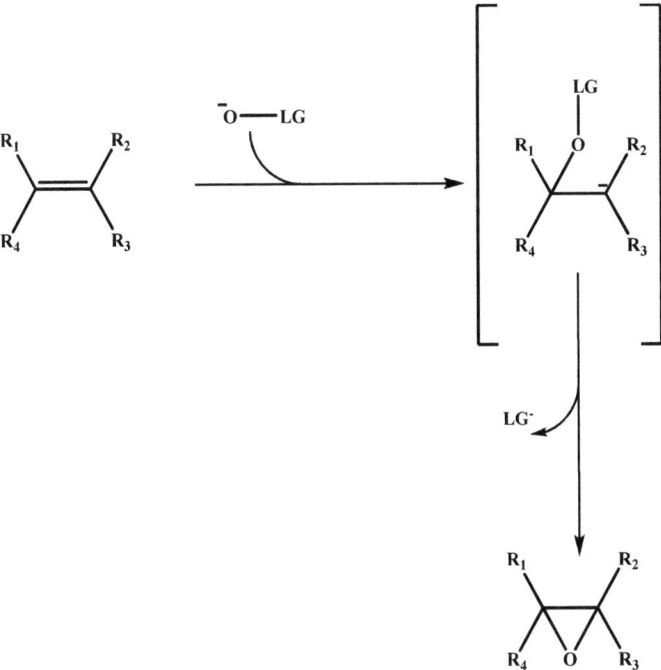

Figure 4: Epoxidation via nucleophilic attack with R_1, R_2, R_3, R_4 being electron deficient groups.

This mechanism requires electron-withdrawing groups (R_1, R_2, R_3, R_4) on the olefin in order to make the double bond "nucleophilic" enough to be epoxidized. Metal alkoxides (usually based on Zn, Mg or La) are usually used in basic media for these kind of reactions in combination with negatively charged alkoxides attacking the C=C double bond[4e,4f].

1.3 Choice of the Oxidation Agent

A wide range of oxidants has been applied in epoxidations, *e.g.*, nitrous oxide[32], iodosylarene, hypochlorite, hydrogen peroxide, organic peroxides, peracids, ozone and molecular oxygen[4d]. Amongst those only N_2O, H_2O_2, O_2 and ozone produce harmless side products such as water, nitrogen or molecular oxygen. Molecular oxygen usually requires an additional reducing agent since four electrons are needed to reduce both oxygens in O_2. Moreover, molecular oxygen is difficult to be activated selectively. This is not only because of its strong oxidation potential and four-electron-reduction, but also because of its triplet ground-state known to be active in autoxidation[5-22] at moderate temperature. This can cause complex side-product formations.

Another example for a suitable oxidant is nitrous oxide with its only by-product being harmless molecular nitrogen. Unfortunately its high oxidation power causes some selectivity issues especially at temperatures above 300 °C as it can produce ketones, aldehydes or even CO and CO_2 from propene or small cyclic alkenes[24,25,32]. Additionally, N_2O is a severe greenhouse gas and its high safety regulations make handling difficult. Nevertheless, oxidations with N_2O are used *e.g.*, by BASF to form cyclic ketons from olefins[28].

H_2O_2 is another potential candidate as an oxidant for epoxidation but has several economic and chemical challenges, which need to be kept in mind. First of all, the production of hydrogen peroxide is rather expensive which limits its usage to the fine chemical sector or to the final part in the production chain where a high value is added per reaction step. This issue is partially solved by economy-of-scale production, *e.g. via* the Solvay process[29]. Secondly, the efficiency of hydrogen peroxide is usually lower than the one for organic peroxides due to decomposition to molecular oxygen. One intermediate of the decomposition is singlet oxygen ($^1O_2$ or $^1\Delta_g$), which can lead to several side reactions *e.g.*, cycloadditions and the *ene* reaction (*Schenk* reaction). The latter is producing a hydroperoxide functionality in α-position to the new double bond as one would expect for autoxidation as well. These *in-situ* produced organic peroxides are actually used for selective epoxidations as in the SMPO process[33]. Even though certain metals[30] catalyze the decomposition of hydrogen peroxide, the real mechanism of singlet-oxygen reactions are poorly understood. Moreover, due to catalyst or solvent interactions, a high degree of quenching of the singlet oxygen makes the use of hydrogen peroxide economically invaluable.

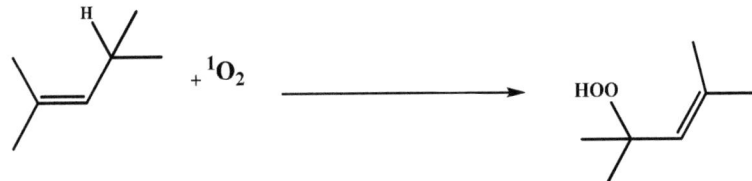

Figure 5: The *Schenk* or *ene* reaction.

Hydrogen peroxide can also be activated heterolytically for the epoxidations of olefins rendering it a versatile oxidant with different activation mechanisms. A famous example is the industrial epoxidation of propene with titaniumsilicate1 (TS-1) and aqueous H_2O_2[2], which is also able to hydroxylate phenol or convert cyclic aldehydes to oximes. The unprecedented activity of this system is due to the substitution of certain silicon atoms by Ti in the MFI structure, a zeolite with channel sizes of 5.5 Å (figure 5).

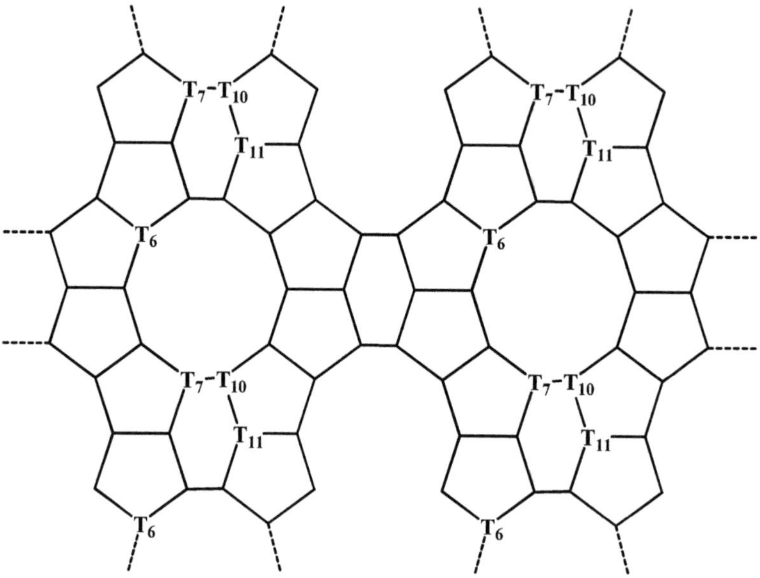

Figure 6: MFI structure with preferred positions for Ti substitution.

Therewith it was possible to force the titanium into a very reactive tetrahedral coordination sphere rationalized by a blue-shifted absorption maximum of LMCT (O→Ti) bands below 230 nm. Even though TS-1 has been the subject to a vast number of scientific publications, the active species is still a matter of discussion. How is the titanium incorporated? Is it bound via one, two or more oxygens to the surface? Does this make a difference for reactivity and stability? Does it make the pores of TS-1 more or less hydrophilic? Is the reactivity due to the tetrahedral coordination or to the very specific chemical milieu which the pores are creating around the Ti atoms? Why can other zeolites which are incorporated with titanium not achieve similar results? Is it possible to create a Ti-containing zeolite with bigger pores for a larger substrate scope while retaining its activity?

Another point that needs to be explored is the decomposition of hydrogen peroxide since it leads to the formation of O_2, which can be explosive upon a certain concentration. This can be a safety problem. Moreover, decomposition reduces the efficiency of the rather expensive oxidant. Therefore, the exact bivalent nature of the active sites in TS-1 needs to be explored in order to reduce hydrogen peroxide decomposition to a minimum, while maintaining the activity in the desired oxidation reaction. *E.g.*, is the same site in TS-1 causing both reactions, decomposition and oxidation? If so, how can the Ti center be chemically tuned to selectively conduct only one of the two reactions? If not, is it possible to synthesize a material with only the wanted active center? Because of the mentioned problems alkyl hydroperoxides are often used for oxidations. These alkyl peroxides produce a stoichiometric amount of the corresponding alcohol, which needs to be separated after reaction and disposed appropriately. Thereby the energy-consuming up-stream separation and the expensive disposal of waste products keep the research field of Ti-mediated epoxidation in the focus of the scientific community.

1.4 Homogeneous Catalytic Epoxidation

In the early 20th century some transition metal oxides were used for the dihydroxylation of olefins with H_2O_2 in *tert*-butanol[34]. In order to be able to dissolve the metal oxides, strongly acidic solutions were used which is why the researching scientists did not observe the epoxide intermediate at first. After switching to neutral or even basic conditions ring opening of the epoxide was prevented. In the 1950ies, vanadium pentaoxide was used as a catalyst with an organic hydroperoxide under neutral to basic conditions[35]. After the discovery of successful epoxidation with V_2O_5 different kinds of early transition metals were tested towards epoxidation activity with alkyl hydroperoxide[36]. Oxides of Mo, Ti, W and V were demonstrated to be the most active. But the activity is also very dependent on the nature of the olefin to be epoxidized (*e.g.*, unfunctionalized *vs.* olefins with hydroxyl functionalities) and in the following decades several trends for successful transformations were found (Table 1)[36]. Because early transition metals are very strong Lewis acids, the more electron-rich the olefin is, the higher is the reaction rate. The reason for that is a stronger overlap of the empty d-orbitals of the metals and the occupied π-orbitals of the double bond. Moreover, *cis*-alkenes were preferably epoxidized to *cis*-epoxides indicating that a similar olefin activation mechanism is valid for early transition metals. Another crucial point was the polarity of the solvent, which should be kept as low as possible since high polarity has tremendous negative effects on selectivity and activity. The decrease in catalytic performance is due to strong solvent-metal interactions, which prevent the coordination of the reactants even though small amounts of *in-situ* formed alcohol were shown to counterintuitively increase the reaction rate[36]. Undesired decomposition of peroxide, which is undesired for reasons mentioned previously, was observed at rather high reaction temperatures around 100 °C causing side product formations through radical chemistry.

Metal	Lewis aciditiy	e⁻-rich olefins	unfunctionalized olefins	Polar solvent	small addition of alcohols
Mo	+++	++	++	-	+
W	++	++	++	-	+
V	+	++	-	--	+++
Ti	+	++	-	--	++

Table 1: Influence of different parameters on the epoxidation rate of Mo, W, V and Ti.

After the discovery of these metals being active in epoxidation reactions, the research field expanded to other metals making it difficult to keep track. In order to provide an overview, the following review is structured by the different metals (or molecules) investigated as catalysts. The review focuses on early transition metals and briefly mentions other metal-based and non-metal-based epoxidations for the purpose of completeness.

1.4.1.1 Molybdenum-based Catalysts

The first active Mo-complexes for epoxidations was patented by Halcon[37]. In this process, dissolved molybdenum oxide proved to be most active for terminal olefins. They are more difficult to epoxidize because of the lower electron density exhibited by the double bond. The oxidant of this reaction is an alkyl hydroperoxide (*iso*-butane or ethylbenzene), which is formed *in-situ* by the autoxidation with molecular oxygen. Large quantities of either *tert*-butanol or 1-methylbenzylalcohol are produced which are further processed to MTBE or styrene[38]. At the beginning of the 1970ies, Mimoun *et al.* proposed a mechanism for the epoxidation with molybdenum oxide in which the olefin is activated by the metal through an interaction with the π-electrons of the C=C double bond[39]. The authors used different additives amongst which HMPT (Tris(dimethylamino)phosphinoxid) proved to have the best performance enhancement of the molybdenum catalyst. Later on, the effect of nitrogen containing ligands coordinated to the active metal was investigated in the 1970ies by Arakawa *et al.*[40]. Sobeczak and Ziolkowski showed that the two main advantages of adding HMPT were increased solubility or substrate binding[41] and a longer life-time[42]. Sheldon and Van Doorn achieved yields up to 64% with $Mo(CO)_6$ within 20 hours for otherwise rather inert terminal olefins (*e.g.*, 1-octene)[43]. However, the same catalyst (as well as $Mo(acac)_6$) did not only convert cyclohexene to over 90%, but showed increased selectivities as well. The same authors supported the common idea of electron-rich olefins being more active by conducting epoxidations with dienes bearing a terminal and an internal double bond. Sheldon and Van Doorn were able to selectively epoxidize the internal double bond over the terminal one. Sheng and Zajacek investigated the reaction with conjugated olefins which were slightly less active[44]. Mo(TPP)OCH$_3$ selectively epoxidized *cis*-2-hexene over *trans*-2-hexene and even small complexes such as hexacarbonyl molybdenum showed a preference for the less sterically demanding *cis*-isomer[42, 43]. The effect of steric hindrance is also very important when a di-substituted double-bond is to be epoxidized in the presence of a tri-substituted. Tri-substituted olefins are sterically more demanding and therewith bulky molybdenum catalysts preferably transfer the oxygen from alkyl peroxides to the less substituted alkenes[42, 45]. The selectivity is reversed when a catalyst with smaller ligands is used[46]. Mo-catalysts were also used for epoxidations of allylic alcohols. One of the first to study Mo/TBHP/allylic alcohol systems were Yamada *et al.* and Hanamoto *et al.*[47, 48]. The catalyst, Mo(O)$_2$(acac)L with L being an achiral hydroxylamine, proved to enantioselectively epoxidize allylic alcohols but suffered from ligand replacement and degradation due to the harsh oxidizing conditions. In order to build more stable complexes a different metal had to be used which interacts stronger with the ligand (V or Ti which will be discussed below). In general, the HO-functionality is thought to have a directing effect. This effect is more pronounced for V or Ti whereas Mo oxide complexes with HMPT selectively epoxidized the allylic double bond in geraniol. However, only small selectivites were observed for the different double bonds in linalool[49]. *Cis*-substituted allylic alcohols produced higher amounts of the *threo*- over the *erythro*-isomer when Mo was chosen as the catalytically active metal[50]. In order to explain the high selectivities (over 90%), the author rationalized the observed results by a preferred coordination of the allylic alcohol to Mo. A right angle was found to be ideal for Mo-mediated epoxidation of allylic alcohols. Recently, Neuenschwander et al. reported on the epoxidation mechanism of peroxo-Mo species[12b]. Computational methods and EPR spectroscopy suggested that the oxygen transfer step and the formation of radical oxygen species is mediated by Mo^{VI}. Based on these results the chemistry taking place during Mo-mediated epoxidation is purely conducted by Mo^{IV} Lewis acid sites without the involvement of redox-chemistry.

1.4.1.2 Tungsten-based Catalysts

As mentioned earlier, the first tungsten based homogenous catalysts used for epoxidations were tungsten oxides dissolved at low pH. Sheldon and Van Doorn[43] used hexacarbonyl tungsten in combination with organic peroxides, building "Mimoun-type" peroxo-tungsten oxides, in a similar procedure as the Mo equivalents, but observed a worse performance of these systems. Surprisingly, the very same systems are used in industry in combination with H_2O_2[36c] for the large-scale production of glycidol. The advantage of using tungsten as catalytic metal is based on the low decomposition rate of hydrogen peroxide therewith forming less active oxygen species responsible for decomposition or side products[36d,51]. In order to epoxidize water-insoluble olefins, a two-phase-system is usually used with an acidic tungsten (or molybdenum) compound as a phase transfer catalyst (PTC) in the presence of a pyridinium chloride salt, thus achieving yields up to 80%[52]. Another example of a biphasic system is given by Venturello et al.[53] who used a phosphate as a co-agent in contrast to a pyridinium salt. Terminal olefins with different substituents were efficiently catalyzed with conversion of over 90% and selectivities ranging from 70-90%. With a similar system, Sato et al. obtained perfect selectivities and good yields for the epoxidation of electron-deficient alkenes, *e.g.*, unsaturated ketones[54].

1.4.1.3 Vanadium-based Catalysts

Vanadium (as well as titanium) mediated homogeneous epoxidation of unfunctionalized olefins is much slower compared to Mo or W mediated ones. Especially terminal olefins usual give very low yields when soluble acac-complexes are used[43], although a few high-yielding terminal olefin examples are known in literature. Mimoun and co-workers achieved moderate epoxidation yields of *cis*-olefins by using a vanadium oxo-complex with a Schiff base and TBHP as the oxidant[55]. Another research group showed epoxidation of cyclohexene with vanadium tartrate complexes but observed low selectivities due to radical side product formation[56]. The main products formed during the reaction turned out to be cyclohexanol and cyclohexenone, which are usually formed by H-abstraction in allylic position upon which molecular oxygen can attack the formed radical. This kind of aerobic reaction conditions in combination with AIBN was first used by Kaneda *et al.* for the direct synthesis of epoxy alcohols from cyclic olefins. Further developments of the aerobic vanadium systems resulted in selectively epoxidizing the C=C double bond[57]. Regarding the epoxidation of alcohols, vanadium catalysts have superior activity over Mo or W based ones when polar solvents are carefully prevented in the reaction mixture. In order to minimize the effect of water or *tert*-butanol (*in-situ* formed from TBHP) by displacing the ligands on the vanadium, different compounds were tested for their stable coordination under reaction conditions[58]. Sharpless *et al.* successfully epoxidized several complex diastereomeric molecules to the *syn*-isomer with TBHP using a VO(acac)$_2$ catalyst[59]. The same formation of *syn*-products was discovered for uncatalyzed epoxidations with peracids[60] but with lower conversions. Sharpless later suggested a perfect angle of 50° around the O-C-C=C bond explaining the high-stereoselectivities of this system[61]. However, VO(acac)$_2$ gives erythro-hydroxy epoxides when cyclic homoallylic alcohols are used as substrates with yields up to 98%[59]. Around the millenium, different researchers investigated epoxidation systems using vanadium and hydroxamic acids with unprecedented enantioselectivities (75-99%), the highest ones achieved with a sterically very demanding hydroxamic acid[62].

1.4.1.4 Titanium-based Catalysts

Until the 1980ies, Mo, W and V have been in the focus of the epoxidation community. This was changed when Katsuki and Sharpless[63] verified that Ti(O-^{i}Pr)$_4$ forms stable complexes with enantiomeric tartrate ligands, which are very active. From then on, selective Ti-mediated asymmetric epoxidation of allylic alcohols was given a new name: "Sharpless epoxidation". Its unexpected high enantiomeric excess (*ee*) of mostly over 90% of the *erythro*-product made it one of the most applied catalytic steps in synthesis of sugars or other alcohol involving target molecules[64]. The epoxidation system consists of an oxidant (TBHP), a readily available and cheap metal precursor (titanium *iso*-propoxide) and a ligand (tartrate). The active complex is formed upon fast replacement of the propoxide by the chiral tartrate ligands due to their higher hapticity. TBHP is used as an oxidant because it is very active even though it has a bulky group attached, which simultaneously helps to induce stereoselectivity because it is sterically more demanding than *e.g.*, H_2O_2. Another crucial point is that TBHP can be prepared in water-free solutions. The exclusion of water is of utmost importance as it increases the reaction rate[65]. Sharpless *et al.* showed in an experiment that the ee drops below 50%[66] after intentionally adding one equivalent of water. Moreover, because Ti is a strong Lewis acid, which are capable of triggering various (cyclic) rearrangements, the reaction temperature should be kept as low as possible in order to minimize this effect and to increase the selectivity in general[67]. The enantioselectivity of the catalytic Ti system is so remarkable that it is possible to selectively epoxidize one enantiomer in the presence of the other, resulting in a reaction mixture with only one enantiomer being epoxidized making post-synthetic work-up easier. This process is called kinetic resolution and it was shown to produce "enantiomerically pure" product formation provided that the ratio of the epoxidation rates between the two enantiomers exceeds 25[68]. During process optimization, the best ligand was identified to be di-*iso*-propyl ester, which for best results needed to be added in a small excess (120 mol%) in order to avoid small amounts of achiral complexes. Sharpless and co-workers proposed different mechanistic aspects regarding the uniqueness of titanium for this system[69]. Ti^{4+} is a d^0 metal with the ability to covalently bind four ligands, two binding sites for the tartrate ligand, one for the peroxide and the last one for the allylic alcohol. Latter ligand is activated by the Lewis acidity of the titanium enabling the nucleophilic attack of the distal oxygen of TBHP. In contrast to the Jorgensen-catalyst or catalysts based on Mo or Fe-porphyrins, the Sharpless system works best with *(E)*-alkenes, minimizing steric interactions[70]. Due to the huge range of applications of this reaction only a few examples are mentioned in here. Walker *et al.* showed that carbonhydrates can successfully be synthesized from different aldehyde precursors which are transformed into an allylic alcohol by a Wittig reaction[71]. In the synthesis of terpenes, which inherently exhibit a lot of double bonds, Sharpless epoxidation is used to introduce oxygen[72]. The Sharpless epoxidation is also part of the synthesis of pheromones for different insects as comstock mealybug[73], mosquitos[74], southern pine beetle[75] or the gypsy moth[76]. The asymmetric epoxidation was also applied to homolytic alcohols achieving lower *ees* but giving the opposite enantiomer[76]. Apart from the typically known Sharpless asymmetric epoxidation, Adam and co-workers demonstrated the direct synthesis from higher substituted olefins to epoxy alcohols by singlet oxygen[77]. Worthwhile mentioning is also the successful epoxidation of rather electron-deficient hydroxylenones by titanium *iso*-propoxide with tartrate ligands and TBHP. The diastereoselectivities obtained with those multiple functionalized substrates exceeded all expectations (>99%)[78]. Additionally, the same Sharpless system can also be used to oxidize sulfur compounds emphasizing its versatile application[67b]. Another very recent discovery is the use of hydrogen peroxide in asymmetric epoxidations in combination with with Ti salan[79a-e] or Ti-salalen complexes[79f,79g]. Katsuki reported the synthesis of a Ti-salen complex which was able to epoxidize conjugated double bonds in very good yields and good to excellent ees. Even for 1-octene good ee ver observed after 24 h of reaction with H_2O_2 (30wt% in water) and were able to detect dimeric peroxo complex which served as a reservoir for the active species[79d]. Exchanging the titanium by the rather expensive niobium was demonstrating to epoxidize allylic alcohols in the presence of aqueous hydrogen peroxide[79e]. Berkesssel et al. synthesized Ti salalen complexes where one imine group is reduced to the corresponding amine in contrast to the original salen ligand[79f]. The authors reported good yield and excellent ees for the asymmetric epoxidation of bicyclic conjugated olefins with aqueous H_2O_2 (30wt%) at room temperature after 18h. Introducing phenyl-groups into the α-position to the phenol groups of the salalen increased the yield and *ees* whereas introduction of electron with-drawing groups were disadvantageous. However, only low yields and moderate ees were observed for aliphatic and monocyclic substrates regardsless of the number of substituents on the double bond. In a different study, the authors reported the deactivation of the active peroxo Ti salalen complex due to oxidative degradation of the amine group based on ESI-mass spectroscopy[79g].

1.4.1.5 Rhenium-based Catalysts

After Rhenium was found to be active in epoxidations, investigations and publications on this area skyrocketed in the late 20th century. Especially methyltrioxorhenium (MTO) proved to be applicable for a wide range of alkenes. Due to the huge amount of Re catalyzed epoxidations, several reviews have already been published on this topic[80]. The ability of rhenium to activate molecular oxygen triggered filing of several patents for the industrial epoxidation of propene[81]. Amongst the ability of activating molecular oxygen, the numerous papers on MTO mediated reactions are also due to its ready availability and high activity at different temperatures, even in the presence of water[82]. Since a significant amount of dihydroxylated products were observed, the addition of nitrogen bases (*e.g.*, pyridine or 3-cyanopyridine) proved to be an efficient method to suppress ring-opening and enhance the selectivity[83]. However, an excess amount of the nitrogen base was necessary in order to balance its own decomposition. Oxidants without water such as urea hydroperoxide or bis(trimethylsilyl) peroxide improved the selectivity[84]. MTO epoxidizes *cis*-olefins only twice as fast as *trans*-olefins[85]. An internal double bond is usually preferred over the terminal one but Copéret *et al.* reported the successful epoxidation of terminal olefins[86] as well. Only a small directing effect by the HO-functionality of allylic alcohols was detected and no significant rate increase was reported[87]. The active species of MTO is thought to be a mono- or bi-peroxo methyl rhenium oxide[88] with a coordinated solvent (or pyridine) molecule. The σ^*(O-O) orbital of these η^2-peroxo-ligands on the rhenium and the HOMO of the C-C double bond play a major role in the transition to the epoxide. While MTO is quite stable towards water exposure, these peroxo species are quite moisture sensitive[89], which explains deactivation in the presence of water[90]. The cleavage of the methyl rhenium bond is a main parameter in this deactivation process and it turns out that under basic conditions nucleophilic attacks on the C-Re bond occurs, deactivating the catalyst[91].

1.4.1.6 Fe- and Mn-based Catalysts

In contrast to early transition metals, iron and manganese complexes epoxidize olefins through a high-valent metal oxo species, meaning that the oxidant first transfers an oxygen to the metal (usually in the center of a very planar bulky ligand like porphyrin or salen) which is further transferred onto the olefin in a subsequent step[92]. Because of this simple oxygen transfer to the metal, a broader range of oxidants can be used for epoxidation, *e.g.*, iodosylarene[93], hypochlorite[94], potassium monopersulfate[95], magnesium monoperoxyphthalate[96], periodate[97], ozone[98], oxaziridines[99], organic peroxides[100], peracids[101] and hydrogen peroxide[102]. The latter oxidant usually requires the use of a basic co-catalyst in order to control the pH-value and avoid oxidant decomposition. Molecular oxygen can also be used as an oxidant, but it is then necessary to add a reductant like boranes[103], metals[104] or ascorbate[105]. In contrast to the Sharpless epoxidation system, the planar metal-oxo catalysts obviously prefer *cis*-alkenes, minimizing disadvantageous steric interactions of the substituents (mainly porphyrin, salen or triazacyclononane). In the beginning, strong deactivation of the metal porphyrins was observed due to dimerization of two porphyrins via a bridging oxygen atom[106]. A method to prevent this dimerization was to attach electron-withdrawing[107] or bulky[108] ligands onto the porphyrin, additional axial coordination of the metal center[109] or even supramolecular encapsulation[110]. Researchers reported all kind of active porphyrin ligands which are usually named according to their geometry, *e.g.*, picnic basket[108b], picket-fence[111], twin-coronet[112], chiral wall[113] or strapped porphyrins[111]. Every synthetically design of these large porphyrin compounds has its own catalytic feature, *e.g.*, Collman's picnic basket (Fig. 5) shows an incredible high selectivity of linear over cyclic olefins or di- over tri-substituted olefins[108b].

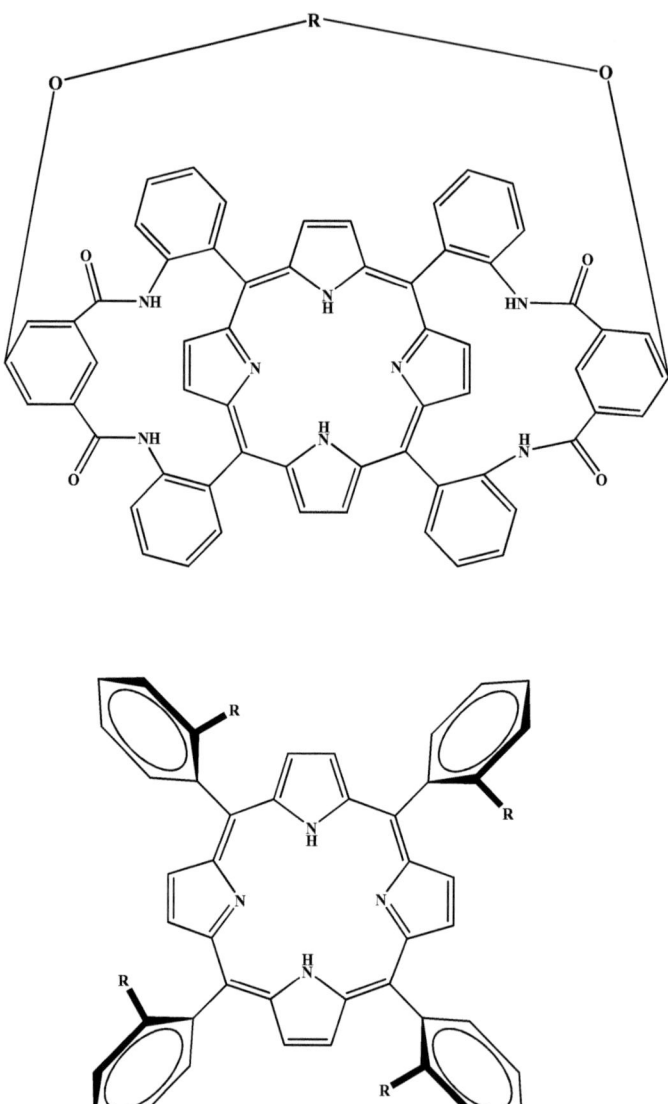

Figure 7: Collman's picnic basket porphyrin (top) and the strapped porphyrin (bottom).

When looking at the picnic basket, it is obvious that only the "inner side of the basket" is selective in epoxidation. The other unselective or open side needs to be blocked by bulky ligands. The acceleration of the reaction rate and enhancement of the selectivity by a fifth, axial ligand was reported by other researchers as well[114]. Pyridine and imidazole were usually the ligands of choice and needed to be added in excess due to oxidative decomposition.

A second class of Mn epoxidation catalysts are built with the help of salen ligands. Jacobsen[115] and Katsuki[116] independently developed a chiral salen Mn(III) catalyst for the formation of optical active epoxides. While Jacobsen only used the diamine backbone to induce stereoselectivity[117], Katsuki also included the 3,3'-positions of the aromatic rings for system optimization[118]. The preferred oxidants were commercially available bleach (NaOCl) or iodosylbenzene, operating in acetonitrile or dichloromethane. Similar to the porphyrin systems, Jacobsen and Katsuki reported the best results for *cis*-alkene preventing unfavorable alkene-salen interactions (side-on approach). However, a Katsuki-type catalyst[119] with two bi-phenyl ligands in the 3,3'-position was reported to slightly alter the rigidity of the complex. Upon oxidation to the active Mn(V) oxo structure this salen ligand folds in a way that the steric interactions of the substituents of the *(E)*-olefins pointing towards the Mn complex is decreased. In this way, the epoxidation of *(E)*-ß-methyl-styrene with *ees* over 90% was achieved. Electronic factors were pointed out by Jacobsen who showed that electron-rich substituents stabilize the electrophilic high-valent transition state of the catalyst[120]. However, the reaction mechanism is still under debate[121]. It was suggested by a group around Seebach and Adam that the reaction follows different mechanisms[122] based on the product ratio resulting from different counter ions and oxidants. Moreover, Mukaiyama demonstrated successfully the use of molecular oxygen for asymmetric epoxidation[123]. Under oxygen atmosphere and with *neo*-pentyl aldehyde as a co-catalyst it was suggested that molecular oxygen forms a peracid with the aldehyde, which subsequently activates the salen Mn(III) complex to the high-valent Mn(V) complex by an oxygen transfer. Using pivaldehyde the same research group was also able to enantioselectively epoxidize different naphthalenes in the presence of *N*-alkyl imidazole as a fifth axial ligand (Fig. 6). Omitting the imidazole leads to the formation of a manganese peroxo complex with low yields and different selectivities. In another study[124] the authors reported opposite enantioselectivities with the same system under aerobic conditions, suggesting that the epoxidation followed the peroxo metal route. A similar effect was discovered by Jacobsen when he added ammonium salts stemming from cinchona compounds[125]. *Cis*-ß-methylstyren is preferably (10:1) converted to the corresponding *cis*-epoxide while the *trans*-epoxide was obtained upon the addition of an ammonium salt.

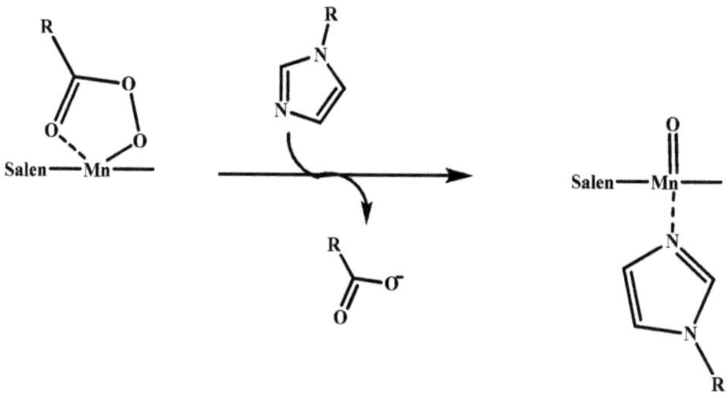

Figure 8: Different active manganese oxo and peroxo species[123,124]

A similar class of Mn catalysts active in olefin epoxidations are ß-ketoiminato complexes. Nagata and coworkers reported aerobic asymmetric epoxidations with these complexes in combination with a co-reductant[126]. Surprisingly, they observed the reverse selective outcome of the reaction when using NaOCl as an oxidant pointing towards two different reaction mechanisms[126a]. Similar to the ketoiminato-complexes, triazacyclononane ligands have also been used to create planar ligand coordination of the Mn atom. The system proved to be highly effective in the epoxidation using hydrogen peroxide in buffered methanol at an pH of 8, but was very active in the before mentioned undesired decomposition of the oxidant[127]. The decomposition could either be minimized by working in acetone, in which the peroxide was "stored" as a hemiacetal[128] or by working in an oxalate buffer[129]. In the same paper, the authors also reported the epoxidation of electron-deficient olefins in moderate yield and terminal double bonds with almost perfect selectivities and reaction rates. Berkessel and Sklorz verified ascorbic acid to be a valuable co-catalyst for this catalyst[130].

1.4.1.7 Non-metal-based Catalysts

In addition to metal-catalysts, it needs to be mentioned that several organic substances are active in epoxidations as well. Amongst those are dioxiranes which are formed upon reaction of a ketone such as acetone and a peroxo compound[131]. The resulting dioxiranes were highly active in epoxidation and yields of over 95% of the *syn*-epoxide were obtained. Making the dioxirane more electron-deficient by fluorinating one methyl group, enhanced the reaction rate tremendously and broadened the substrate scope[132]. Denmark *et al.* studied the epoxidation of different ketones in biphasic systems[133]. Molecular engineering of the ketone revealed an oxopiperidinium structure, which showed the highest activity in dioxirane formation in a suitable pH range slightly over 7. Together with the slow addition of potassium peroxomonosulfate (OXONE), undesired decomposition or other side reactions were decreased to a minimum. In addition, they also proved that the epoxidation mechanism involves dioxirane formation by ^{18}O labelling. Using a less flexible ketone Kurihara and coworkers demonstrated the successful epoxidation of a range of terminal olefins[134]. Because trifluoromethylacectone is not very sterically demanding, (Z)-disubstituted alkenes have been epoxidized in high yields[135] in contrast to the systems developed by Jacobsen and Katsuki. Nevertheless, the reaction rate was slower when HO-functionalities were close to the C=C double bond. However, Wang was successful in selectively epoxidizing several olefins with a chiral ketone with a fructose body[136] even in the presence of functional groups such as esters or acetals.

Another class of organic compounds active in homogeneous epoxidation are iminium or ammonium salts. Wynberg and Marsman were the first ones reporting the use of quinine and ephedrinium salts as epoxidation catalysts[137]. Another group of researchers improved the system by using TBHP and *N*-benzylcinchonidinium chloride in toluene with good *ees* and moderate yields[138]. Aggarwal and coworkers reported the epoxidation of styrene with ammonium and iminium salts with OXONE in high yields but low enantioselectivities[139]. Even simple amine salts such as triethylamine were active. (S)-2-diphenylmethypyrrolidine epoxidized different cyclohexenes with over 90% yield and moderate *ees*. Nelson reported the use of ammonium salts as a phase transfer catalyst in toluene and water with moderate reaction times up to 24 h[140]. Bortolini *et al.* used an iminium salt based on binaphthylamine with different olefins[141]. The rate of reaction increased as expected from the mono-, over the di- to the trisubstituted alkenes.

Johnson *et al.* reported the formation of epoxides from a mixture of sulfides, chloroalkanes, an aldehyde and a strong base[142]. The reaction involves the formation of a sulfonium ylide with the chloride from the chloroalkane as the counter ion. The alkyl chain on the chloroalkane is subsequently electrophilically attacked by the aldehyde thereby forming an epoxide from the alkyl chain of the chloroalkane with the alkyl chain of the aldehyde. Enantioselectivity can be introduced either by the substrate or by using a chiral sulfonium ylide[143a]. By using a Ru or Co salt it was possible to increase the formation of the chiral sulfonium ylide and to achieve incredibly good yields and enantio- and diastereoselectivities for *trans*-stilbene oxide[143b].

Another area of metal free epoxidation systems with H_2O_2 is the use of slightly acidic organic compounds in order to create an appropriate chemical environment for the activation of oxygen transfer to the olefin but without triggering hydrogen peroxide decomposition[144]. Jacobs and co-workers, *e.g.*, reported the successful epoxidation of various di- and trisubstituted olefins in a mixture of phenol (pK_a=9.9) and aqueous H_2O_2 (50wt%)[144a]. They achieved good to excellent yields and ees in less than 24 h at 60°C. Electron with-drawing groups in *meta-* or *para-*position in phenol accelerated the reaction rate whereas the addition of proton acceptor groups as tetrahydrofuran decreased it. Berkessel and Adrio reported acceleration of cyclooctene epoxidation by 100 000 times through the addition of 1,1,1,3,3,3-hexafluoro-2-propanol (HFIP, pK_a=9.3) and suggested a highly ordered transition state due to an experimentally oberserved activation parameters (*e.g.* $\Delta S^\neq$= 39 J · mol^{-1} · K^{-1})[144c]. While the kinetic orders in hydrogen peroxide and the substrate were 1 the kinetic order in HFIP varied between 1 and 12 depending on its concentration[144b]. The authors suggested that a HFIP coordination sphere of about 12 HFPI molecules is responsible fot the enormous rate acceleration comparable to hydrogen-bonding networks in proteins. Moreover, the epoxidation of conjugated olefins with a peptide catalyst consisting of three aspartate molecules was demonstrated by Berkessel[144d]. Good yields but low ees were obtained for at room temperature with hydrogen peroxide. However, decreasing the reaction temperature to -10°C increased the ees to over 80% proving that the hydrogen-bonding network of peptides can epoxidize olefins with large polar groups in the homoallylic position at low temperature. In another study, Berkessel et al. synthesized a dendritic polymer with hexafluoro-hydroxy-groups at the end of the polymer branches[144e]. Good yields with H_2O_2 (50wt%) were obtained for trisubstituted olefins and cyclohexene at 40°C in dichloromethane after 24 h of reaction emphasizing the versatility of hydrogen-bonding networks. However, it should be noted that monosubstituted olefins were only obtained in low yields.

1.4.1.8 Other Catalysts

For the sake of completeness a few more examples will be given, which were demonstrated to be active in homogeneous catalysis. Ruthenium is another metal active in epoxidations although its inherent strong oxidation power can cause C=C bond cleavage which would result in aldehyde formation in the presence of an oxidant[145]. The use of nitrogen containing ligands helped to alter the electronics of Ru in a way to decrease C=C cleavage and to increase the selectivity towards epoxidation[146]. A porphyrin Ru complex was reported by Groves and Quinn to activate molecular oxygen for epoxidation but due to slow reaction rates and disproportionation of the active monooxo Ru species its performance is worse than the one reported for porphyrin systems with Fe or Mn[147]. The above mentioned Jacobsen- and Katsuki-catalysts based on salen were first used by Kochi *et al*. with Cr instead of Mn[148]. Active in epoxidations with iodosylbenzene or pyridine-*N*-oxide, they were even able to isolate the (salen)Cr^V-oxo intermediate. Salen ligands were also used in the combination with Co as a center metal with molecular oxygen as the oxidant[149]. There are reports on Cu- and Ni-mediated epoxidations[150].

1.5 Conclusions

After over half a decade of research on homogeneous catalytic epoxidation scientist have developed tools to selectively functionalize a wide range of alkenes with good to very good yields and outstanding selectivities. Depending on the class of substrate to be epoxidized the right catalytic system can be found in literature. Terminal olefins can efficiently be epoxidized by MTO with hydrogen peroxide and 3-cyanopyridine as mentioned above[86]. For disubstituted double bonds either the Jacobsen[115] system or the Katsuki[119] Mn salen with biphenyl ligands in the 3,3' position can be applied depending whether a *cis*- or a *trans*-olefins is the substrate. $Mo(CO)_6$ and $W(CO)_6$ can be used for the epoxidation of higher substituted substrates in order to avoid unfavorable steric interactions[43]. Of course, allylic alcohols are preferably converted by the Sharpless epoxidation system which does not only result in high *ees* but due to the precise molecular insight into the mechanism one can easily predict the produced enantiomer by choosing the right tartrate ligand[63]. All these specific chemical solutions verify the huge achievements made in this research area. However, even though mechanistic proposals have been suggested for the different systems, some reaction steps are still controversially debated. *E.g.*, it is still unclear why some catalysts show different selectivities when the oxidant is changed from molecular oxygen to NaOCl[126]. The same effect is observed when Jacobsen added the cinchona alkaloid derived salts to his reaction conditions[125]. Apparently, two different reaction pathways are present giving different product selectivities upon changing the catalytic metal from Ti or V to Mo as it has also been observed by Rossiter, Verhoeven and Sharpless[50].

Even with deep mechanistic insight into homogeneous epoxidations, those systems still bear an inherent problem. The work-up after conversion of the alkene in order to separate the catalyst from the liquid reaction solution is rather energy intensive. Therewith, catalyst recovery is often a cost intensive issue. In order to solve this problem, heterogeneous systems with similar selectivities are investigated in cases for which molecular structures were more difficult to obtain due to the lack of analytical tools. By heterogenization of the catalytic species, spectroscopists are often confronted with new unfavorable surface effects not observed in homogeneous solutions. The reaction mechanism is complicated due to additional effects (*e.g.*, adsorption or desorption). Hence, it is more probable to contribute to the multi-disciplinary area of heterogeneous epoxidation catalysis than to the field of homogeneous epoxidation.

1.6 Heterogeneous Catalytic Epoxidation

In order to overcome the problem of energy intensive catalyst recovery several attempts were made towards immobilizing transition metals active in olefin epoxidation on organic, hybrid or inorganic supports.

The first category of supports is usually a polymer with strongly coordinating groups which can bind the target transition metal. The linkages between the monomeric units of those polymers need to be stable towards oxidative degradation under reaction conditions. Moreover immobilization on a support always goes along with the risk to introduce additional active sites in oxidant decomposition, epoxide opening or other side reactions. Thus polyesters exhibiting acidic terminal groups inherently are a rather bad choice. But due to a wide range of possible functional groups which can be introduced into a polymer all kinds of metals are able to be introduced. However, strong metal-polymer interactions need to be provided in order to avoid leaching of the catalytic species[151].

The second class are catalytic hybrid materials which consist of an inorganic support covalently bound to an organic linker whose other end coordinates the active metal or covalently binds it as well[152]. In principle, all kind of metals can be immobilized depending on the right linker which can be synthesized accordingly[153]. The inorganic support is usually chemically and thermally stable enough under the reaction conditions[154]. However, the organic linker and its two connections (to the support and the active metal) need to be stable in the oxidative milieu as well, bearing the same problem as mentioned before for a polymer support[151]. Additional side product formation can occur on the inorganic support or by terminal groups of the organic linker, which did not take up a transition metal.

The last method mentioned is the direct fixation of metals onto (via post-synthetic grafting) or into (incorporation during or after synthesis) a purely inorganic support. The latter is restricted to metals which can be forced into a tetrahedral coordination sphere in order to replace a framework silicon[155]. Titanium, vanadium and chromium are metals which are stable under those molecular restrictions and are therefore good candidates for incorporation into zeolites or other silica supports. Metals of higher groups or columns usually prefer a higher coordination number limiting their immobilization on inorganic supports to post-synthetic grafting. This method can be conducted in the gas (Chemical Vapor Deposition = CVP) or in the liquid phase. Anchoring of the metal precursor [MX_n] is usually done upon reaction with an acidic surface HO-group (*e.g.* silanol groups) yielding a new surface metal species [$\equiv SiO\text{-}M(X)_{n-1}$] and a leaving group [H-X]. The precursor is usually a cheap metal chloride or alkoxide but various other ligands building stable leaving groups can be used in the liquid phase as well. When grafting is done in a solvent it is important to wash the solid after reaction and to confirm the complete removal of leaving groups, excess precursor and solvent in order to avoid any interference of these materials during epoxidation. Additionally, it is also difficult to obtain absolutely moisture-free solvent in order to avoid hydrolysis of metal-support or metal-ligand bonds. The interference of water is a problem often encountered in epoxidation[65,66].

The CVP method is however a very convenient way to prepared surface anchored metal species under absolute moisture-free conditions and without the use of any solvent reducing the amount of waste. In this synthesis method, a volatile precursor is transferred in an excess via the gas phase onto the support under high vacuum (< 0.1 mbar). The high volatility necessary for chemical vapor deposition restricts the selection of applicable precursors to metal carbonmonooxides or chlorides as $TiCl_4$, $VOCl_3$ and CrO_2Cl_2, which are known to have melting points below room temperature. The complete removal of the excess metal compound and the leaving group can be achieved by a treatment under high dynamic vacuum (0.01 mbar) and low temperature. However, due to water adsorption on dehydrated silica and fast hydrolysis of metal chloride bonds it is necessary to store the materials in a glove box under moisture-free conditions (<1 ppm water).

1.6.1 Molybdenum-based Catalysts

Sherrington and co-workers reported the successful immobilization on different polymers with *N*-containing functional groups[156]. However, the heterogeneous Mo catalyst suffered from leaching and low mechanical stability. Kotov and Boneva investigated polymer supported $MoO_2(acac)_2$ (amongst other catalysts containing Ti or V) which showed moderate to good yields in the epoxidation of different olefins a slight leaching. However, the obtained yield and reactions rates were very dependent on the initial concentration of TBHP and did not exceed its homogeneous equivalent[157]. Ziólkowski and co-workers investigated a similar system using μ^3-oxo trimetal precursors (M=V, Cr, Mn, Co, Ru, Rh, Mo) heterogenized onto polymers[158]. The molybdenum oxide showed the highest conversion of cyclohexene and lowest decomposition in cumyl hydroperoxide which was used as an oxidant. Thomas et al. grafted MoO_2Cl_2 in tetrahydrofuran onto MCM-41. The homogeneous compound and the heterogeneous counterpart showed very similar kinetic results in yield and reaction rate proving that surface grafting did not inactivate the complex[159]. However significant amounts of leaching were observed and thus raising the question if the active species is truly heterogeneous. Gonçalves and co-workers investigated the same system with 1 wt% Mo. In the presence of triethyl amine higher loadings were grafted but amongst leaching and epoxide ring opening poorer TOF per Mo were obtained[160]. Nevertheless, good to excellent conversions of octenes and norbornene were reported for MCM supported MoO_2Cl_2 with perfect selectivities towards the epoxide after 6h. Loadings below 1wt % Mo haven't been prone to leaching and showed good recyclability. Using (N≡)Mo(O^{t}Bu)$_3$ as a precursor on silica and MCM-41 resulted in heterogeneous catalysts with loadings up to 6.7 wt %[161]. Depending on the pretreatment temperature bipodal [(SiO≡)$_2$Mo(≡N)(O^{t}Bu)] or monopodal surface complexes [(SiO≡)Mo(≡N)(O^{t}Bu)$_2$] leaving the Mo≡N triple bond intact as proven by IR and NMR spectroscopy. The monopodal catalyst was very active in epoxidation of aliphatic terminal olefins with conversions around 73% and 90% or higher for higher substituted olefins. Hot filtration test proved the heterogeneity of the process but strong deactivation was observed in the first few runs. A different study reported the synthesis of porous siliceous materials with molybdenum oxide in the pore channels[162]. The heterogeneous system was active in anhydrous reaction media with TBHP and showed good (for limonene) to excellent (for cyclooctene) conversions at 40°C within 3 h. Gonçalves' research group tried to immobilize Mo complexes organic linker but the obtained structures were prone to either leaching, epoxide ring opening or deactivation[163]. Kühn and co-workers observed low activities of grafted η^5-CpMoCl(CO)$_3$ due to diffusion limitations in H-ß and H-Y[164]. Using zeolites with larger pores circumvented this problem and TOF similar to the homogeneous counterpart were reached and the surface Mo species did not leach if no organic linker was used[165]. However, deactivation was observed when the surface complex was linked via a carbon backbone which the authors attributed to coke or other organic molecules adsorbed on the surface. No linkers were used by Maiti and co-workers a molybdenum peroxo precursor for on MCM-41 with excellent conversions and good selectivities[166]. Their system was superior compared to the homogeneous catalyst which is probably due to the use of a H_2O_2/$NaHCO_3$ reaction mixture which built up peroxy carbonates keeping the real hydrogen peroxide concentration low and therewith avoiding its decomposition. M. Masteri-Farahani[167] et al. reported the synthesis of Jacobsen-like and Schiff-base Mo complexes with excellent selectivities and TOF up to almost 60 h^{-1}. The research group of Thiel reported that a η^2-butyl peroxo Mo species was activated in order to transfer the distal oxygen onto the olefin[168]. In a different study, they were able to directly incorporate ligands into the MCM-41 structure and post-synthetically introduce $MoO(O_2)_2$. The material showed excellent epoxidation yields, selectivities and stability[169]. When iron oxide particles were used as a support for more convenient post-reaction separation[170]. Despite of lower reaction rates, the differently supported complex was still very active and selective. Don Tilley et al. grafted $MoO[OSi(O^tBu)_3]_4$ and the tungsten analogue to a SBA-15 surface[171]. Even though the activities of the complexes were rather low TBHP showed almost perfect selectivities towards the epoxide while the use of hydrogen peroxide triggered side product formation and oxidant decomposition. Some systems were reported in the direct oxidation of propene with molecular oxygen using

Ag/MoO$_3$/NaCl compositions but conversions were below 2%[172]. However, Song et al. grafted K$_2$MoO$_4$ to a silica surface and epoxidized propene in the gas phase with molecular oxygen[173]. They achieved considerable conversions of 14% with over 30% selectivity towards the epoxide at 300°C with acetal aldehyde being the main side product.

1.6.2 Tungsten-based Catalysts

Tungsten has been used in many PTC systems but only a few examples of heterogeneous catalysts are known. Sherrington did also immobilize W but with the same stability problems as for molybdenum[156]. In another study, he and his co-workers used polyglycidylmethacrylate as support and grafted tungsten on it with the help of triamides attached to the polymer. No leaching or ring opening were observed and TOFs up to 1235 h^{-1} were reported with molecular oxygen as oxidant[156g]. Hammond and co-workers reported very high activity of a W/Zn/SnO$_2$ material obtained by flame spray pyrolysis. The material did not suffer from leaching, was stabe during three runs and showed an increased activity and selectivity in cyclooctene epoxidation compared to a benchmark catalyst. Surprisingly, calcination of the material resulted in a decrease of the activity and triggered leaching of the active species into the reaction mixture. Sels, De Vos and Jacobs reported the immobilization of monomeric and oligomeric peroxotungsten compounds on layered double hydroxides[175]. The material was very selective in terms of epoxide formed but TON were only acceptable as well as hydrogen peroxide selectivity. The same group also used organic linkers with phosphoramide groups at the distal end to bind tungsten[176] similar to the linkage groups Sherrington used for tungsten complexation. Conversion were moderate to good within 24 h and the selectivity towards the epoxide and hydrogen peroxide. No leaching was observed but the material suffered from serious epoxide ring opening as a side reaction. Tungsten phosphoramide complex which were directly grafted to the MCM-41 surface without organic linker showed lower activity. Jacobs and co-workers also reported an ion-exchange amberlite by the Venturello anion $\{PO_4[WO(O_2)_2]_4\}^{3-}$ being active for the epoxidation of different terpenes[177]. High epoxide selectivities and conversions after two or three days were achieved with hydrogen peroxide at 38 or 0°C. Some substrates were epoxidized in biphasic media with no to moderate epoxide formation and the addition of (aminomethyl)-phosphoric acid promoted the reaction rates. Mizuno et al. immobilized dimeric tungsten diperoxide species on silica. They used charged imidazolium rings linked to the silica surface by a carbon backbone with the dimeric tungsten anion as a counterion[178]. Good yields and with H_2O_2 in acetonitrile were achieved and the catalyst was recovered up to three times without the loss of activity. Don Tilley used SBA-15 supported $WO[OSi(O^tBu)_3]_4$ for epoxidations with TBHP and H_2O_2 as mentioned before[171]. In 2006, Grivani and co-workers showed that a polystyrene supported tungsten hexacarbonyl is active in combination with hydrogen peroxide in acetonitrile. The catalyst proved to be stable and could be used up to ten times in the epoxidation of cyclooctene[179]. Recently Gao et al. dispersed WO_3 species onto a mesocellular silica foam with good stability and yields in the epoxidation of cycloocta-1,5-diene with hydrogen peroxide[180]. They rationalized based on spectroscopic studies that mainly isolated tungsten sites are responsible for the formation of epoxide.

1.6.3 Rhenium-based Catalysts

After Herrmann et al. reported a convenient method to synthesize MTO[92] they patented a method to immobilize MTO on *N*-containing polymers a year later[181]. But the patented heterogeneous system only exhibited low activities. The activities were increased with good selectivities by using a pyridine cross-linked polystyrene support which resulted in an octahedrally coordinated Re oxocomplex with two pyridines from the support[182]. The prepared heterogeneous systems were not only active with hydrogen peroxide they were also stable up to at least 5 runs. Adam and co-workers observed an effect of the hydrogen peroxide selectivity based on the zeolite pores in which MTO was anchored[183]. The increased selectivity in combination with the achieved stability of the solid catalyst made it an interesting candidate in olefin epoxidation. Espenson and colleagues successfully demonstrated the epoxidation of different soybeans derived oils with MTO supported on niobia in water-free reaction media with urea-hydrogen peroxide[184]. They observed conversions above 70% within 2 h for different olefins at room temperature in chloroform. The heterogeneity of the system was proven and the catalyst was reused several times without the loss of Re. They were also able to show the superior catalytic performance of the heterogeneous system over the homogeneous and proposed two different bipodal peroxo rhenium surface species to be responsible for epoxide formation. A combination of polymer and inorganic support was chosen by Neumann et al. They grafted polyethers or polypropylene on silica and absorbed the MTO on this hybrid material[185]. They only observed moderate to good selectivities at 25°C with aqueous hydrogen peroxide (30wt%) as an oxidant but the catalyst was not stable over more than five runs. Nunes et al. linked a bipyridyl ligand to MCM-41 to which ReO$_3$ was coordinated in a subsequent step with a simultaneous shrinking of the pore diameter. However, EXAFS analysis disclosed a major part of the ReO$_3$ which was not coordinate by bipyridyl ligand in the active epoxidation catalyst[186]. Kühn and co-workers grafted acylperrhenates onto different zeolites and tested their performance in cyclooctene epoxidation with aqueous H$_2$O$_2$[187]. They reported better conversion for the supports bearing no aluminum due to lower Lewis acidity. The catalyst suffered from strong deactivation and leaching. The latter problems were also observed by Gelbard and co-workers who reported the immobilization of MTO onto aminated polystyrene, polyacrylate and vinylpyridine[188]. The MTO was coordinated to the nitrogens in the polymer support as rationalized by FTIR, UV-VIS and ^{13}C NMR spectroscopy and proved to be active in α-pinene epoxidation with H$_2$O$_2$.

1.6.4 Iron- and Manganese-based Catalysts

Rosa and co-workers successfully encapsulated iron-porphyrins into NaX zeolites and tested these for different oxidations[189]. In the epoxidation of cyclooctene with iodosylbenzene the materials performed well with yields ranging from 85 to 95% and hydroxylation was observed when adamantane was used as a substrate. The authors attributed the latter observation to a radical formation similar to cytochrome P-450 initiating a radical chain mechanism. In a different study by the same authors they compared the iron with manganese porphyrins and claimed the first being more active in oxidation of based on the observed product distribution[190]. Zhang et al. anchored the porphyrin ring directly to the MCM-41 wall enabling substrate approach from both porphyrin faces[191]. TOFs up to 196 h^{-1} within the 2h of reactions were reported with good to excellent selectivities to cyclooctene oxide. The authored excluded leaching and thus confirmed the heterogeneity of their system. Zhan and Li tried to immobilize Mn porphyrin derivates in faujsite-Y via the "ship-in-a-bottle" approach[192]. They obtained moderate conversions in cyclohexene epoxidation with TBHP and serious ring opening. Overoxidation could be decreased by the addition of pyridine and no leaching or deactivation was observed. Martinez-Lorente et al. were able to oxidize alkanes and epoxidize alkenes with hydrogen peroxide by Mn-porphyrin linked to the pore walls of K10. Yields up to 88% for epoxidation were reported but major side product formation by homolytic O-O cleavage were observed. Battioni and co-workers very stable halogenated iron and manganese porphyrin complexes linked to silica or montmorillonite K10[193]. Epoxidation of cyclooctene with iodosylbenzene gave good yields of 80% cyclooctene oxide or higher. However, radical chain mechanisms forming side products as ketones and alcohols were observed. Tangestaninejad and Mirkhani reported the successful epoxidation of limonene by polymer supported manganese porphyrin in acetonitrile/water with imidazole as co-catalyst and $NaIO_4$ as oxidant[194]. The porphyrin was linked via an ether bond to a short hydrocarbon chain which was linked via oligopeptides to the polymer backbone and were very stable upon recovery and reuse of the catalyst. However, epoxide yields were only moderate to good even after 72 hours. Catalytic porphyrin systems covalently bound to poly(phenylesters)dendrimers were used for the epoxidation with iodosylbenzene of different dienes and alkenes[195]. The authors tested the selectivity of these large systems towards mono-/di- and di-/tri-substituted double bonds and cyclic *vs.* linear olefins. In both cases only the second generation dendrimeric supports exhibited significant selectivities towards the sterically less hindered double bond. Cooke and Lindsay Smith directly compared fluorinated iron and manganese porphyrin complexes covalently linked to poly(vinylpyridine), imidazole modified polystyrene and silica, the latter performing best in epoxidation of cyclohexene with iodosylbenzene[196]. The product distributions were very similar to the homogeneous epoxidations of the corresponding porphyrin complex but again resulted in radical autoxidation. However, the system were not prone to leaching and could be reused. Another approach to obtain polymerized fluorinated iron porphyrin catalysts is to link the different porphyrin rings by octafluorobiphenyl-units[197]. The synthesized catalyst was very active in hydroxylation of cyclohexane but showed only low conversions in epoxidation of norbornene. Turk and Ford reported good conversions of styrene to styrene oxide at room temperature with NaOCl (TON of 620 in 1h)[198]. They used colloidal anion-exchange particles to immobilize Mn porphyrins in their pores of 60 nm diameter. One drawback was the oxidative degradation of the porphyrin ring though. Anzenbacher et al. covalently bound iron- and manganese-porphyrin rings to polystyrene[199]. The catalyst was used to oxidize styrene with a co-catalyst(imidazole/pyridine) and reached epoxide yields between 69 and 90% depending on the metal and co-catalyst used. During the course of reaction the chemical bonds between the catalytic metal and porphyrin centers and the support stayed intact.

Besides metal porphyrins two other Mn complexes have been immobilized on a support. The first group are salen complexes and Salvadori and co-workers were the first who reported their immobilization on polystyrene[200]. They reported very high yields of polystyrene and 2-methyl-polystyrene at 0°C with magnesium monoperoxo phthalate and *m*CPBA which even showed some stereoselectivity. The research group around Sherrington additionally chose polymethacrylate as a support for the Jacobsen's catalyst and achieved conversions up to 49% and e.e.'s between 60 and 91%[201]. Even though no to negligible leaching was detected the catalysts strongly deactivated during two or three runs. De et al. reported the heterogenization of Jacobsen-type complexes onto polymers (ethylene glycol dimethacrylate) with comparable activities to the homogeneous system but the enantioselectivity drop from 80 to 30% upon immobilization[202]. The encapsulation into zeolite Y resulted in conversions between 10 and 40% with good to excellent Epoxide selectivities[203]. Moderate ees were only achieved with sterically more demanding substrates but still comparable to the homogeneous analog and no leaching could be detected. Bein and Ogunwumi epoxidized styrene derivates with Mn salen complexes encapsulated into zeolite EMT[204]. Only low conversions and moderate epoxide selectivities were achieved with NaOCl but by addition of pyridine *N*-oxide conversion was promoted up to 47%. A filtration test underlined the heterogeneity of the system together with the inability of the catalyst to epoxidize cholesterol within 18 h. In a multiple-step synthesis an organic linker was first attached to MCM-41 which was subsequently connected to the first half of the Jacobsen catalyst. Finally the salen ligand was completed and the metal was added by manganese acetate[205]. Very high conversions and ees were achieved at 0°C with reactions times as short as 15 or 45 min. No leaching was detected and after washing and drying the catalyst could be reused up to four times.

The other class of Mn complexes active in enantioselective alkene epoxidation are triazacyclononanes and have been attached to solid materials, too. De Vos and co-workers successfully incorporated these complexes into NaY zeolite and were able to epoxidize cyclohexene and styrene with excellent epoxide selectivities at 0°C and in a H_2O_2/acetone mixture[206]. Amongst oxidant decomposition they observed side products resulting from radical autoxidation but now further reaction after a hot filtration test. The same author also grafted triazacyclononane to silica surfaces via an organic linker[207]. However, in contrast to the homogeneous system, overoxidation and diol formation was observed for all olefins tested. The same catalyst was used by Sels and co-workers for the epoxidation of different terpenes but they also observed the formation of side products by radical autoxidation and rearrangements[208]. Rao et al. prepared a similar material but they linked the Mn complex to the silica surface with a different linker[209]. Good selectivities and low TON are reported and methanol was a better solvent than acetone and acetonitrile. Seger and Janda immobilized the Jacobsen catalyst on different polystyrene or other resins[210]. The systems were as active as the homogeneous counterpart but all catalyst lost their enantioselectivity (and in some cases also their activity) after three runs. Don Tilley and co-workers reported the synthesis of a silica supported iron catalyst which epoxidized cyclooctene with hydrogen peroxide in high selectivities at 25°C[211]. However, the authors observed enone and alcohol for cyclohexene epoxidation under the same conditions.

1.6.6 Titanium-based Catalysts

Heterogeneous epoxidation catalysts based on titanium are widely used in industry for the production of propene oxide with a capacity of 8 Mt/a[2]. 56.8% of which is produced by catalytic process involving a Ti-containing material[212a] while the rest is produced by an old non-catalytic chlorohydrin process based on research from the 19th century[212b]. The four other Ti-mediated processes are Sumitomo's cumene hydroperoxide (CHP)[213], Shell's Propylene oxide/styrene monomer (PO/SM)[214], Halcon's propylene oxide/*tert*-butyl alcohol (PO/TBA)[215] and BASF/Dow's hydrogen peroxide/propene oxide (HPPO) process[216]. The latter process uses a zeolite named Titaniumsilicate-1 (TS-1). As shown in figure 5, the MFI-type zeolite consists of a pore network with a diameter of 5.5 Å spanning a 10-membered ring of SiO_4-tetrahedra[217]. Certain positions in this silica framework are substituted by tetrahedral Ti atoms as studied by neutron powder diffraction[218]. Inexpensive silica supported titanium are used as catalysts for the first three processes, *e.g.* silica grafted $TiCl_4$ in the PO/SM process[33]. A simple sol-gel method based on the reaction of TEOS and titanium alkoxide indicated that Ti atoms in a silica matrix need to be isolated in order to obtain cyclooctene conversion with hydrogen peroxide[219]. Moreover, Baiker and co-workers investigated how the preparation of titania-silica aerogels with supercritical CO_2 influences the performance on propene epoxidation[220]. They reached high activities and selectivities up to 93% to peroxide and 100% to epoxide with temperature from 60-80°C. Jorda et al. grafted isolated titanium fluoride surface species onto a silica support[221]. They epoxidized cyclohexene in the presence of H_2O_2 and pointed out that a drop-wise addition of H_2O_2 is advantageous as it suppresses side products formed by decomposition. Another grafting precursor for the preparation of isolated Ti sites on silica or zeolites is titanocene dichloride $(Cp)_2TiCl_2$[222]. The authors achieved good yields with TBHP as an oxidant but using H_2O_2 led to irreversible catalyst deactivation. Adam and co-workers grafted the same precursor onto MCM-41 or ITQ and tested their materials for allylic alcohol epoxidation moderate to good conversions with high diastereomeric ratios over 95:5[223]. Corma et al. only investigated the activity of layered aluminosilicates for cyclohexene epoxidation[224]. Good conversions and excellent selectivities towards the epoxide were obtained and the catalyst showed no deactivation after three runs. Apart from the W-based polyoxometalates mentioned above, Kholdeeva and co-workers were able to synthesize a polyoxometalate containing both tungsten and titanium atoms[225]. They were able to epoxidize cyclohexene at 40-50°C with high selectivities (epoxide and H_2O_2). From their kinetic and computational studies they proposed a mechanism with a very important proton transfer step to be of major importance for the epoxidation mechanism. Gao et al. used a structurally similar polyoxometalate and demonstrated its high selectivity in epoxidation with H_2O_2[226]. However, oxidant decomposition, small side product formation and low activities did not render the system very attractive for epoxidation reactions. Scott and co-workers grafted titanium *iso*-propoxide onto dehydrated silica surfaces and claimed the formation of dimeric Ti-surface sites[227]. They achieved good to excellent conversions and selectivities in the epoxidation with TBHP at temperatures from 25 – 65°C but did not report on the stability of their catalyst. The research group of Don Tilley did a lot of research on silica grafted Ti catalysts for epoxidation. They observed good TOF with cumyl hydroperoxide for (^{i}PrO)Ti[OSi(O^{t}Bu)$_3$]$_3$ grafted on Aerosil, silicone nanosphere, SBA-15 or MCM-41[228]. The latter two increased the epoxidation activity and calcination reduced it. They also reported epoxidation of cyclooctene with H_2O_2 over SBA-15 supported Ti catalysts with chemically modified silica surfaces[229]. Even though TON were low, oxidant selectivities up to 71% were achieved with a methylated surface modified by fluorinated acetate. A study on a dimeric Ti catalyst by the same group did not show any specific reaction rate increase compared to heterogeneous epoxidation catalysts with isolated Ti sites[230]. Another group reported increased epoxidation rates and selectivities upon silylation of Ti-grafted HMS which increases the hydrophobicity of the catalyst[231]. Another preparation method was used by Can Li et al. which deboronated silica xerogels and grafted $TiCl_4$ onto the material[232]. Unfortunately only low conversions and selectivities were observed with cyclohexanol and cyclohexenone as side-products. Thomas and co-workers grafted $CpTi[Si(O^tBu)_3]$ onto MCM-41 and observed good conversions with high selectivities in epoxide and TBHP[233]. A loading of 2 wt% Ti proved to be the best catalyst and conducting the reaction under argon reduced the selectivities at higher conversions. Corma and co-workers synthesized Ti-Beta zeolites with different Al and Ti loadings and compared their kinetic performance to TS-1 for the epoxidation with H_2O_2. Very low conversions with a lot of side product formation and H_2O_2 decomposition were reported[234]. Moreover TS-1 was verified as the superior catalyst except for the epoxidation of cyclododecene where Ti-Beta was more active. Dealuminated Ti-Beta was considerably more active in terms of TON and TOF with hydrogen peroxide efficiencies up to 94%[235]. The purely siliceous framework provided a hydrophobic reaction environment in which different cyclic and aliphatic olefins were epoxidized with selectivities ranging from 68.3 to over 98% to the epoxide. In agreement with this observation D'Amore and Schwarz reported improved epoxidation yields after silylating TS-1 with N,O-*bis*(trimethylsilyl)trifluoroacetamide[236]. Moreover, Figueras, Kochkar and Caldarelli prepared hydrophobic titaniumsilicates under acidic synthesis conditions and epoxidized cyclohexene at 75°C with hydrogen peroxide[237]. 52% of the converted cyclohexene was epoxidized while the rest reacted to allylic side products. Wu and Tatsumi prepared Ti-MWW materials which selectively epoxidized the *trans*-olefin of a racemic hex-2-ene mixture[238]. High selectivities in oxidant and epoxide were reported and rationalized by the sinusoidal pore system of the MWW structure. Choudary, Valli and Prasad used a titanium-pillared montmorillonite catalyst for the asymmetric epoxidation of several bi- and tri-substituted allylic alcohols between 15 and -20°C[239]. High yields with ees over 90% were reported when tartrate was added to the reaction mixture and no molecular sieve as water scavenger was needed. Polymerizing tartrates and 1,8-cyclooctandiol and impregnating the resulting polymer with Ti(O^{i}Pr)$_4$ results in a catalyst with moderate yields and ees for the epoxidation of olefins with hydroxyl-functionalities at -15-20°C[240]. However, the reaction time (days-weeks) were very long and a four-fold addition of tartrate with respect to titanium atoms were necessary for asymmetric epoxidation.

Decades of research have been done on the famous Titaniumsilicate-1 in order to elucidate the local structure around the Ti atom and how it is bound to or in the framework. The most important achievements will be shortly mentioned here. Due to its Ti loading below 3 wt% Ti obtaining reliable data on the Ti surrounding network was difficult because of the strong background signal from the silica matrix[241]. Ground breaking was the study of Bordiga and co-workers in 1994, where they used EXAFS to prove that the Ti atoms are isolated in the MFI structure and tetrahedrally coordinated[242]. It is no surprise that EXAFS was the first technique to clarify the local structure around the Ti atoms. Due to its element specific excitation it was possible to circumvent the problems resulting from interference of the Si atoms even though relatively long measurement times were necessary, a consequence of the limited Ti loading. EXAFS usually only give an averaged picture of the investigated atom species but due to the chemical uniformity of the different Ti sites in TS-1 useful information could be obtained for this system in contrast to others elements as Fe and

Ga[243]. The findings from Bordiga et al. were supported by other groups using different spectroscopic techniques supporting the hypothesis that Ti is incorporated into Si-vacancies in the MFI structure. IR and Raman spectroscopy showed new signals upon Ti incorporation at 960 cm^{-1} [244] and 1125 cm^{-1} [245]. Additionally UV-Vis spectroscopy reveals a signal around 220 nm (48 000 cm^{-1})[246,247] which is significantly blue-shifted compared to commercial Ti samples as anatase or rutile, both exhibiting signals over 300 nm. Moreover, an increasing unit cell of the TS-1 structure was observed upon Ti incorporation by powder diffraction [248]. Neutron powder diffraction was finally the technique which provided a sufficiently large contrast between Ti and Si atoms to determine the preferred sites in the MFI structure for Ti incorporation[218]. Further evidence that TS-1 is a material with a high number of defective site and the successive incorporation of Ti atoms was given by DTG data[249]. The authors investigate the condensation of silanol nest in the temperature range between 500 to 800°C and monitored the mass decrease. From those data they obtained an inversely proportional correlation of vacancy sites against Ti incorporated. To a similar conclusion came Millini et al. when they observed a decreasing number of silanol protons by ^{1}H MAS NMR the more heteroatoms (Ti or B) incorporated in the MFI structure[250]. This correlation points towards the substitution of four silanols for every heteroatom added to the framework which would result in closed (tetrapodal) Ti sites [(≡SiO)$_4$Ti]. However, Lamberti and co-workers observed a coordination number of 4.45 ± 0.25 in the first coordination sphere which is significantly over 4 and therewith indicating the existence of opened Ti site with on Si-O-Ti bond hydrolyzed [(≡SiO)$_3$Ti(OH)(HO-Si≡)][251]. Furthermore, Bordiga et al. showed that the positions T7, T10 and T11 are adjacent to each other which allows to form two Ti sites which differ slightly from each other[252].

T7 and T10 can form Ti-dimers when al positions are occupied or Ti sites with a vacancy in close proximity while T6 forms isolated Ti sites. T11 is a site next to T10 but is not a part of the larger pore system making absorption of molecules on this site difficult. Thus, it is probably not participating in oxidation. Elucidation of the active species in the TS-1 has been subject to a lot of publications. Two structures have been in the main focus of discussion, the η^1-hydroperoxo and η^2-peroxo Ti species[253]. Bonino and co-workers impregnated a dehydrated TS-1 sample with a H_2O_2/H_2O solution and observed a strong decrease of the first and second shell signal along with a color change of the wet TS-1 to yellow[254]. The yellow color disappeared after leaving the catalyst drying overnight and the signal of the first and second shell was (partially) restored. Wetting the catalyst with water again reduced the EXAFS signal again proving the reproducibility of this process. Prestipino et al. first impregnated the dehydrated TS-1 with an anhydrous H2O2 source [$KH_2PO_4 \cdot H_2O_2$] and observed no color change unless water was added to this impregnated sample[255]. The results were interpreted by both groups of the formation of a η^2-peroxo Ti species upon exposing TS-1 to water and hydrogen peroxide along with the cleavage of one or more Ti-O-Si bonds due to the second shell restructuring in EXAFS. However, if the impregnated material dries or is only exposed to anhydrous hydrogen peroxide the η^1-hydroperoxo Ti species is formed and the bonds to the framework stay unaffected. Due to the hydrophobic environment created in the pores of the TS-1 it is speculated that the active species is actually the η^1-hydroperoxo Ti which is present under dry conditions. A lot of literature has been published on the possible reaction which can be catalyzed by TS-1[256]. A more detailed study on the kinetics of TS-1 in propene epoxidation, was made by Baek Shin and Chadwick which found out that the order in H_2O_2 is 0.67 and 0.63 in propene and the activation energy is as low as 25.8 kJ · mol^{-1}[257].

1.6.7 Other Heterogeneous Catalysts

Some other materials do also show activity in olefin epoxidation. Liu et al. were able to immobilize a ruthenium porphyrin complex in MCM-41 and epoxidize a series of bulky substrates which could not be epoxide by TS-1 due to its small pore sizes[258]. Good yields (over 90%) and TOFs (up to 209 h^{-1}) were achieved but the catalyst suffered from leaching and/or deactivation after three runs. However, Ru-porphyrins supported on polymers were reused several times and were demonstrated to be active in olefin epoxidation of similar substrates[259]. TON of 9500 and yields up to 86% were achieved. Very recently Scotti et al. reported the one-pot synthesis of trans-stilbene oxide over a CuO/Al_2O_3 catalyst with high to excellent conversions and very good selectivities of di-substituted olefins with molecular oxygen and *iso*-propylbenzene as a co-catalyst[260]. The catalyst maintained conversions above 90% over three runs and only a slight decrease in selectivity was observed. Another material which was found to be active in epoxidation of a wide range of olefins is silica grafted Ta catalysts reported by Tilley and co-workers[261]. They observed conversions up to 42.7% cyclohexene oxide at 65°C after 2 h with hydrogen peroxide. Main side products were the enone and the allylic alcohol. Selectivities of over 99% were reported for those catalysts after capping the material with hydrophobic surface groups[262]. However, TON after 6h were below 200 and the selectivity based on H_2O_2 was 13.5% or lower. The third biggest organic oxidation process in terms of capacity [18 Mt/a] is the epoxidation of ethane[2]. The reaction is done with molecular oxygen in the gas-phase over an alumina supported silver catalyst[263]. The reaction mechanism is supposed to occur via a radical oxygen species on the silver surface. Similar to ethane epoxidation, the direct oxidation of propene to propene oxide by molecular oxygen is also under investigation[264]. The discovery that this system gives excellent selectivities at 50°C was made by Hayashi and co-workers[265]. The process needs to be fed with molecular hydrogen or a hydrocarbon source which indicates an intermediate formation of hydrogen or alkyl peroxo species[266].

1.7 Conclusions

As in the research area of homogeneous epoxidation catalysis many achievements have been accomplished in the last decades on the field of catalytic materials for heterogeneous epoxidation. Among them, are definitely the environmentally friendly large-scale productions of ethene oxide from molecular oxygen over silver catalysts as well as the HPPO process from BASF based on TS-1 mediated epoxidation of propene oxide with hydrogen peroxide. For the first process a lot of mechanistic insight has already been obtained and the addition of promoters further helped to decrease undesired side reactions. The latter process is not that well elucidated. Even though isolated and tetrahedrally coordinated Ti sites are thought to be responsible for the catalytic activity it is still debated which peroxo species (η^1 or η^2) is transferring the oxygen onto the C=C double bond. Another question is the superior catalytic performance of TS-1 compared to similar zeolites. Is it a pore-size effect or is the hydrophobicity of the pores responsible for the exceptionally high selectivity? Ongoing research on systems producing H_2O_2 *in-situ* from molecular hydrogen and oxygen have shown promising results. However, it seems as the formation of a Ti-peroxide intermediate cannot be circumvented as the direct oxidation of propene over a silver catalyst with molecular oxygen gives low selectivities. Another question which has not been risen so far is whether the substrate is coordinated to the Ti center and if the epoxidation mechanism depends on the oxidant used in the reaction. For instance, if epoxidation with hydrogen peroxide is occurring via a η^2-peroxo titanium species what would be the active species with TBHP? TBHP can bind in a η^2 fashion but only in the peroxo form with the *tert*-buytl group still attached whereas the hydrogen peroxide molecule could be coordinated in doubly deprotonated state. Is it possible that two different species are responsible for the oxygen transfer? Do those have different activation energies for epoxidation?

Part II

Synthesis of Heterogeneous Catalysts

2 Grafting TiCl$_4$ onto Silica

In industry, the epoxidation of propylene is catalyzed by TiCl$_4$ grafted to dehydrated silica. In order to obtain reliable structure-activity correlations the active surface species, which are not yet fully understood, must be elucidated. In this study, a variety of techniques are used to thoroughly follow the speciation and restructuring of site-isolated Ti^{IV} Lewis acid centers. Initially, a ≡SiOTiCl$_3$ species is formed which transfers of Cl-ligands to the silica surface upon heating, leading to multipodal species, e.g. (≡SiO)$_3$TiCl or (≡SiO)$_2$TiCl$_2$. Continuous flow experiments were conducted to demonstrate the superior activity and stability of such multipodal species (i.e., two or three fold bonded to the surface) for catalytic olefin epoxidation.

2.1 Introduction

Industrially speaking, homogeneous catalysts suffer from substantial disadvantages compared to heterogeneous analogous, especially for large-scale processes[159b,267]. Nevertheless, homogeneous catalysts are often well characterized, making structure-reactivity correlations possible. On the other hand, catalytic surfaces often contain a broad range of active species with different selectivity, complicating fundamental studies. This is valid for a variety of industrially applied catalysts. A key example of a catalyst which is poorly understood is $TiCl_4$-grafted silica. This catalyst is applied in the Styrene-Monomer-Propylene-Oxide (SMPO) process for the epoxidation of propylene using ethylbenzene as a oxygen shuttle and molecular oxygen as the oxidant[2,4a,33,228,229,263,268,269,270,271]. Researchers from academia and industry have still not been able to fully characterize the active site, even after decades of investigation. Nevertheless, it is believed that the active titanium species are in a tetrahedral geometry[33,268]. Significant deactivation is observed after several days on-stream and different studies indicate leaching and agglomeration of highly dispersed Ti-species. Agglomeration is thought to take place *via* the formation of soluble titanium complexes, which can be re-deposited to the silica and/or titanium species downstream in the catalyst bed[272]. The resulting growth of the TiO_x particles leads to a decrease in activity compared to the isolated tetrahedral sites. A further decrease is observed due to poisoning by traces of by-products[272].

Dehydration of the silica is the first step in the synthesis of the SMPO catalyst, followed by grafting $TiCl_4$ to the surface silanols in the gas phase. Although monopodal species (*i.e.*, ≡$SiOTiCl_3$, singly bonded to the surface) are often assumed to be formed first, some reports suggest the direct formation of multipodal species[273]. Moreover, the benefical effect of the thermal treatment after grafting $TiCl_4$ is also not fully understood and how it influences the nature and speciation of the Ti surface species. Providing insight into the molecular structure of such catalysts, and improving the performance of the SMPO and related catalysts, is therefore an intellectual and practical challenge.

2.2 Results and Discussion

2.2.1 Grafting TiCl₄ to isolated silanol sites

Heating Aerosil200© (from Evonik) up to temperatures of 700°C under high dynamic vacuum, results in the condensation of hydrogen-bonded silanol groups, without a change in its morphological properties, *i.e.* porosity or surface area. As a result isolated surface silanols are formed, characterized by a sharp IR-signal at 3745 cm^{-1}. ^{1}H MAS-NMR, as well as IR spectroscopy, shows that the average silanol surface-density decreases from approximately 2.5 nm^{-2} at 200 °C, to 0.7-0.9 nm^{-2} at 700 °C[266,274]. Silica supports pretreated at 700°C are often used for the synthesis of single-site catalysts[159a,275,276]. Even though precise nature of the surface silanols is not fully established, it is generally believed that the isolated ≡SiOH groups formed at 700°C are attached to fairly unstrained SiO-rings (*i.e.*, 6-membered rings or larger)[277]. The inherent constrains of surface-species must be contained by an appropriate cluster for the computational investigation of the reactivity of such single-site species, along with some flexibility associated with known physicochemical properties. Small clusters as in Figure 1 can model complex surface species due to a rigid tris(siloxide) assembly featuring a hydrogen-bonding network between the three siloxides. The hydrogen-bondings maintain Si-O-Si angles similar to those assumed in dehydrated silica. Cluster calculations have successfully been used to model similar systems[278]. The scaled harmonic O-H stretching frequency of 3750 cm^{-1} of those clusters agrees well with the experimental value of 3745 cm^{-1}.

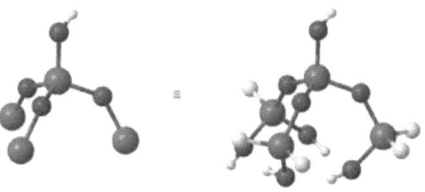

Figure 1. Small cluster used for the computational description of the ≡SiOH sites.

Grafting of TiCl$_4$ leads to the formation of HCl and can be observed by *in situ* gas phase IR spectroscopy *via* its characteristic rotational-vibrational manifold at 2900 cm^{-1}. The implied reaction between the isolated ≡SiOH and the TiCl$_4$ can be followed by a nearly quantitative disappearance of the ≡SiOH signal (Figure 2). ICP measurements of the material (heated to 50 °C) reveals a Cl/Ti ratio of 3±0.1, in line with ≡SiOTiCl$_3$ species. The formed HCl during reaction does not react with silica below 175 °C as show by Haukka *et al.*[279] under dynamic vacuum at room temperature. Physiorbed TiCl$_4$ remaining after thermal post-treatment (T$_{post}$ = 50 to 450°C) can be excluded, based on the observed hydroxyl site stoichiometry and the Cl/Ti ratio being lower than 4, as well as the UV-Vis spectra reported in figure 5. A very small amount of ≡SiOH groups did not react with TiCl$_4$ as can be observed in Figure 2. The remaining silanol groups are most likely inaccessible. However, migration of those ≡SiOH groups cannot be excluded leading to the formation of new siloxane bridges and water molecules.

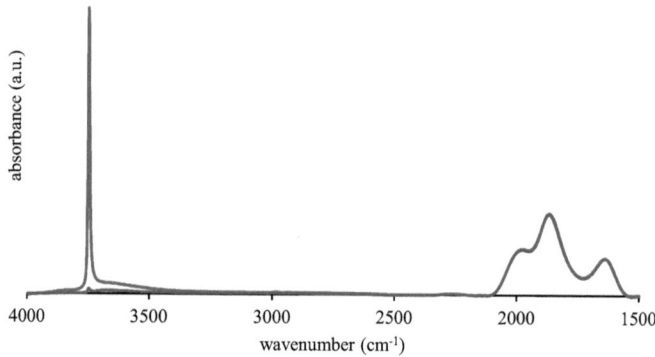

Figure 2. Transmission IR spectra of silica pre-treated at 700 °C, before (blue), and after (red) TiCl$_4$-grafting (T_{post}=50 °C).

The reaction of TiCl$_4$ and isolated silanol groups was computationally investigated and found to proceed *via* a pre-reactive complex (see Figure 3). The decreased adiabatic barrier for this complex was only 6.0 kcal mol^{-1}. Subsequently, HCl forms a post-reactive hydrogen-bonded complex, before its elimination *via* a variational transition state (entropy driven). Due to the average distance between the Ti-species being more than one nm, lateral interactions between the surface ≡SiOTiCl$_3$ sites seem unlikely.

Based on these experimental observations and computational predictions, it is assumed that TiCl$_4$-grafting results in the formation of isolated ≡SiOTiCl$_3$ sites. This hypothesis is, however, in contrast to a literature report claiming that TiCl$_4$-grafting would lead to multipodal sites, even at room temperature[269]. Isolated ≡SiOTiCl$_3$ sites are also in disagreement to recent studies reporting that the grafting of trimethylrhenium and triethylaluminum[280] can yield dinuclear species, grafted either *via* two terminal or two bridged siloxy ligands. The formation of multipodal Ti-species will be addressed below.

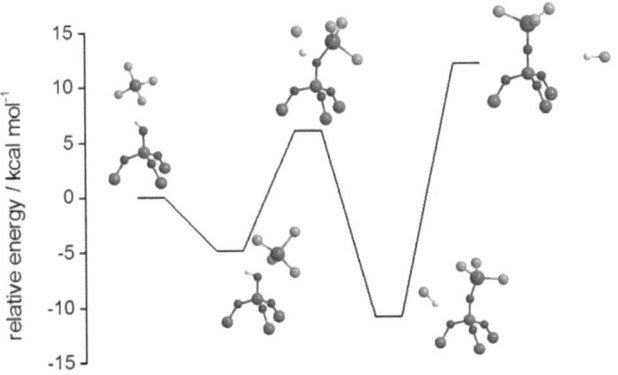

Figure 3. Adiabatic potential energy surface for the grafting of TiCl₄ to an isolated ≡SiOH site.

After grafting of TiCl₄, the SMPO catalyst is typically heated to 900 °C at which post-treatment temperature it exhibited the best propylene oxide yield[281]. The restructuring taking place during this thermal treatment is, however, not fully understood. Ti-loading remains constant at around 1.3±0.05 w% up to 200 °C, but thereafter decreases rather rapidly to 0.8±0.05 w% at 500 °C. The Cl/Ti ratio qualitatively follows the same trend as the Ti-loading, fluctuating from 3.0±0.1 up to 250 °C, followed by a rapid decrease to 2.0±0.1 at 500 °C (Figure 4).

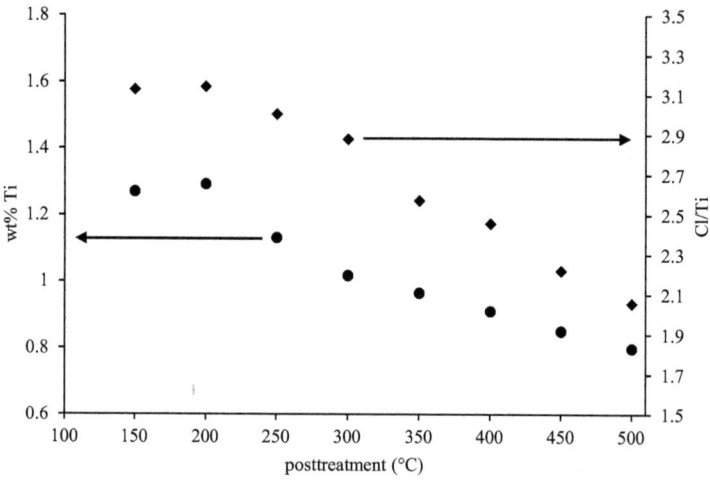

Figure 4. Cl/Ti ratio (♦) and Ti-loading (●) as a function of the post-grafting treatment temperature T_{post}.

Above 200 °C, the fundamental vibration of TiCl$_4$ at 500 cm^{-1} can be observed in the gas phase by transmission IR spectroscopy (glass cell with KBr windows; see Figure 5). Whilst accounting for the decrease in Ti content, the thermal elimination of TiCl$_4$ was unexpected given an average distance between two Ti^{IV} sites of more than 1 nm. Direct interaction of two Ti-species seems unlikely and therewith the elimination TiCl$_4$ without intermediate restructuring.

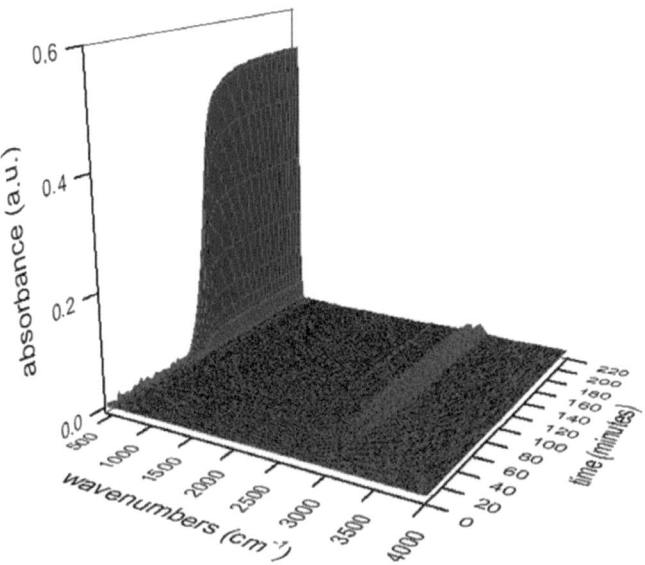

Figure 5. Gas phase IR spectra showing the formation of TiCl$_4$ (monitored at 500 cm^{-1}) and HCl (2900 cm^{-1}) during thermal treatment; ramp rate of 3.5 °C min^{-1} (20 to 500 °C).

2.2.2 Spectroscopic and computational investigations

The ligand-to-metal charge transfer transitions of the Ti^{IV}-species were investigated with diffuse reflectance UV-Vis spectroscopy (Figure 6)[282]. A clear shift of the peak maximum can be obserbed from 260 over 250 to 230 nm for samples treated at 50, 250 and 450 °C, respectively, indicating a significant change in the surface speciation which requires a more detailed study.

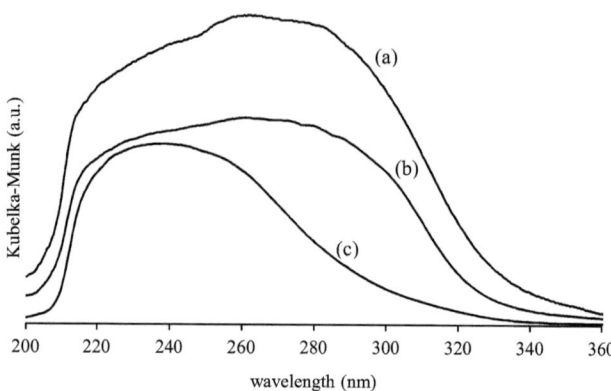

Figure 6. Diffuse Reflectance UV-Vis spectra of the different Ti-SiO$_2$ materials, heated to 50 °C (a), 250 °C (b) and 450 °C (c).

NMR spectroscopy is a powerful technique to identify molecular compounds, even when they are of inorganic nature[283-284]. In figure 7, the change in static ^{35}Cl-NMR powder spectra of TiCl$_4$-grafted samples with thermal post-treatments can be followed at different temperatures T$_{post}$ (see experimental and computational methods). The spectrum of the sample heated to 50 °C (upper trace of Figure 7) shows a rather small signal which is characterized by a small quadrupole coupling constant (QCC) of ≈4 MHz (based on a simulated fit of the powder spectrum). The spectrum of the sample post-treated at 250 °C (middle trace of Figure 7) contains contributions from an additional signal featuring a significantly broader QCC on the order of 12.5 MHz. For the sample heated to 450 °C (lower trace of Figure 7), the narrow component was not observed.

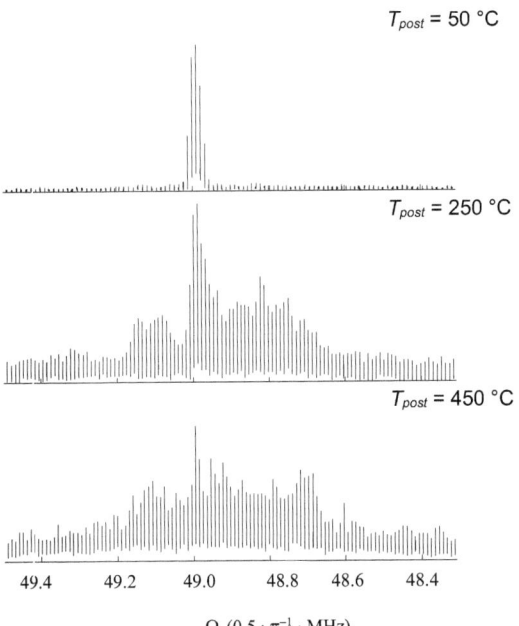

Figure 7. Static ^{35}Cl-NMR spectra of materials treated at 50, 250 and 450 °C (T_{pre} = 700 °C); recorded on a 500 MHz spectrometer. The sharp and intense peak at 49 MHz in the sample post-treated at 450 °C is an experimental artefact and should be neglected.

In order to correlate the observed signals in the ^{35}Cl NMR powder spectra with possible surface species, quantum-chemical calculations were used to predict the QCC of monopodal, bipodal and tripodal Ti^{IV}-species (see Figure 8). Static QCCs of 13.7 (average of the three Cl-atoms of the monopodal species), 14.5 (average of the two Cl-atoms of the bipodal species) and 15.0 MHz were calculated assuming rigid atom position without any molecular motion[285]. Unlike the bi- and tripodal species, the monopodal species is able to rotate around the Si-O-Ti bond, leading to an improved calculation of a dynamically averaged QCC of 5.7 MHz. The assumption of a fast rotation, averaging out the special orientation of the quadrupolar tensor is justified as the rotational barrier is computationally predicted to be as low as 0.2 kcal · mol^{-1}, *i.e.* << R*T*. These predicted QCC values support our hypothesis that TiCl$_4$ grafted to silnaol initially forms a rotating monopodal ≡SiOTiCl$_3$ species. Furhtermore, ^{35}Cl-NMR measurements disprove the claim that multipodal Ti species are formed in the reaction of TiCl$_4$ with dehydrated silica at room temperature[273]. The significantly larger QCCs observed for samples treated at a higher temperature suggest that Ti-species lose their ability to rotate around the Si-O-Ti bond and restructure to more constrained species.

Using the time-dependent DFT method, UV-Vis absorption spectra of models for mono-, bi- and tripodal species (Figure 8) were predicted. Comparing these predictions (Figure 9) with the experimentally obtained UV-Vis spectra in Figure 6 suggests a gradual restructuring of the surface species from ≡SiOTiCl$_3$ to (≡SiO)$_2$TiCl$_2$ and (≡SiO)$_3$TiCl.

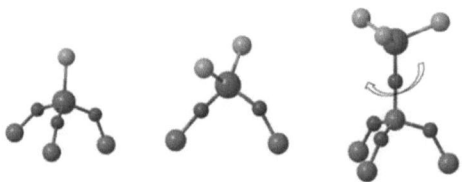

Figure 8. B3LYP-DFT optimized models for ≡SiOTiCl$_3$, (≡SiO)$_2$TiCl$_2$ and (≡SiO)$_3$TiCl surface species.

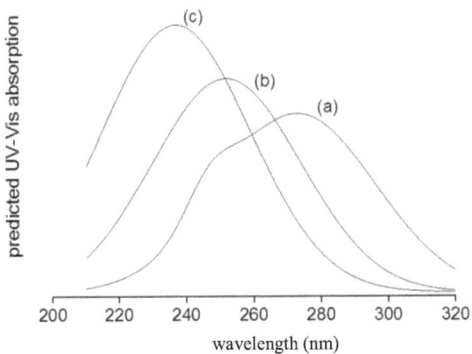

Figure 9. Computationally predicted UV-Vis spectra of ≡SiOTiCl$_3$ (a), (≡SiO)$_2$TiCl$_2$ (b) and (≡SiO)$_3$TiCl (c).

Summarizing the computational and experimental results obtained so far leads to the following intermediate implications: (i) TiCl$_4$-grafting forms ≡SiOTiCl$_3$ species which appear to be stable up to 50 °C as demonstrated by ^{35}Cl-NMR spectroscopy; (ii) heating leads to significant restructuring resulting in multipodal species with larger QCCs in ^{35}Cl-NMR; (iii) above 250 °C, TiCl$_4$ is eliminated explaining the loss in Ti loading; (iv) samples heated to 450 °C show no evidence for remaining monopodal species, due to the missing narrow signal in ^{35}Cl-NMR spectroscopy.

Unfortunately, ^{35}Cl NMR is not able to distinguish between bi- and tripodal species since both species have similar QCCs. The nature of the surface species was therefore probed with N,N-bis-(trimethylsilyl)methylamine (denoted (TMS)$_2$N-CH$_3$). After exposure of the titanium grafted silica samples to this amine probe (see experimental), the liquid nitrogen trapped effluent after exposure was investigated with ^{1}H-NMR. For the materials heated to 50 and 250 °C, the formation of (trimethylsilyl)chloride (TMS-Cl) was detected, indicating a reaction between the Ti-Cl species and the amine probe, observed for Cp*TiCl$_3$ dissolved in methanol as well. For the material heated to 450 °C, TMS-Cl was not observed.

The reaction between the TiCl$_x$ surface species and the (TMS)$_2$N-CH$_3$ was computationally predicted and found to be very exothermic (see Scheme 1), except for tripodal species where $\Delta_r E$ was only -4.7 kcal mol^{-1}.

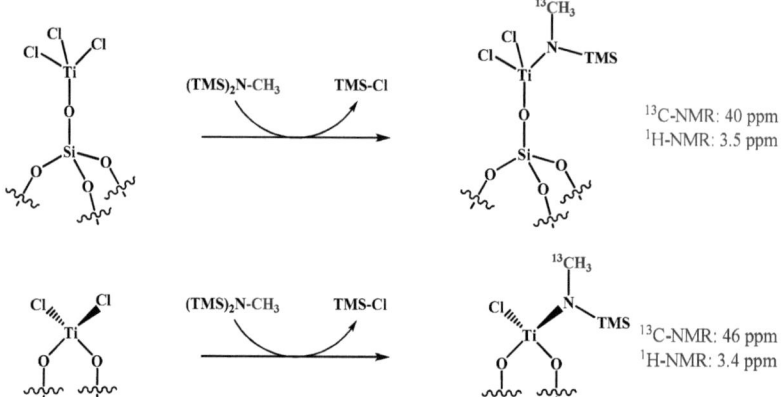

Scheme 1. Reaction of the (TMS)$_2$N-CH$_3$ probe with the mono- and bipodal Ti^{IV} species; reaction energies and predicted ^{13}C- and ^{1}H-NMR shifts for the methyl group in the probe.

The experimental ^{1}H-NMR spectra of the various samples after amine grafting are shown in Figure 10. The methyl proton-signals are split, due to scalar or *J*-coupling with the ^{13}C (see experimental) which leads to a decrease in intensity but to a better signal resolution. It is observed that the sample which was heated to only 50 °C mainly features a signal around 3.7

ppm, with an additional minor contribution at 3.5 ppm. Due to the better resolution with labelled methyl-groups, deconvolution of the two signals could be achieved. The behavior of both signals upon heating is in line with the ^{35}Cl-NMR spectrum of this material (top panel of Figure 7), mainly revealing a signal from freely rotating ≡SiO-TiCl$_3$ species. For the sample heated to 250 °C, the contribution of the peak at 3.5 ppm significantly increased which is predicted to originate from an amine probe reacted with bipodal Ti-species. For the sample heated to 450 °C both signals are absent. The latter observation is in line with the absence of TMS-Cl formation in the effluent (see above). In ^{13}C-NMR (see figure 11), the samples heated to 50 and 250 °C feature a broad peak between 33 and 40 ppm; for the sample heated to 450 °C this signal is also absent. These observations confirm that the amine probe does not react with the species present after heating to 450 °C and hence confirm that those species are fundamentally different from those dominating at lower post-treatment temperatures.

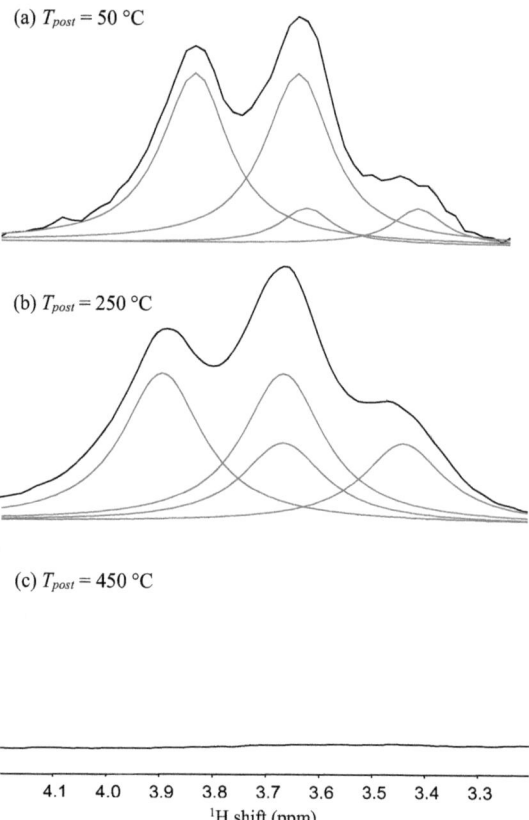

Figure 10. ^{1}H-NMR spectra of the materials treated at 50 °C (a), 250 °C (b) and 450 °C (c), after exposure to the (TMS)$_2$N-^{13}CH$_3$ probe molecule.

Deconvolution and comparison of the experimental spectra with the predicted ^{1}H- and ^{13}C-NMR shifts for the mono- and bipodal species (Scheme 1) supports our hypothesis that the sample heated to 50 °C mainly contains monopodal ≡SiOTiCl$_3$ species (≈85 %). The sample heated to 250

°C shows and increased amount of bipodal species, in line with the contribution of two compounds in the ^{35}Cl-NMR spectrum (middle panel Figure 7). These results can however not exclude the presence of tripodal species in the sample heated to 250 °C, but clearly indicate the absence of monopodal species in the material heated to 450 °C.

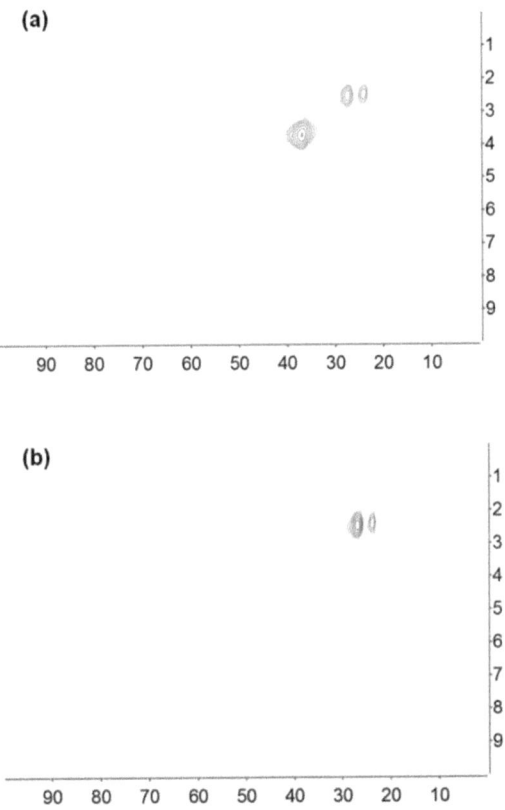

Figure 11. ^{1}H ^{13}C CP-NMR spectra of the materials treated at 250 °C (a) and 450 °C (b), after exposure to the (TMS)$_2$N-^{13}CH$_3$ probe molecule.

The IR spectra before and after amine-probe grafting for the materials heated to 250 and 450 °C can be observed in figure 12. Two strong peaks are visible in the range 3000-2870 cm^{-1} resulting from the proton stretching vibrations of the trimethylsilyl groups. For TiCl$_4$-grafted silica samples heated to 50 °C (not shown) and 250 °C, two additional –CH$_3$ vibrations are identified at 2830 and 2800 cm^{-1}. Those vibrations shift by -8 cm^{-1} when the methyl group on the nitrogen is labelled, while the TMS-peaks at higher wavenumbers stay unchanged. The 2800 cm^{-1} signal corresponds to the symmetric –CH$_3$ stretch of the amine grafted to either mono- or bipodal Ti-species (computationally predicted frequency of 2900 cm^{-1}). In NMR, those protons appear between 3.4 and 3.7 ppm (Figure 10) and the C-atom between 33 and 40 ppm (see Figure 11). The third signal at 2830 cm^{-1} is also present on titanium-free silica samples due to the reaction of the amine probe with siloxane bridges (Scheme 2)[23]. The same can be observed in NMR spectroscopy where those protons appear around 2.5 ppm and the ^{13}C-signal around 27 ppm (see Figure 11). Grafting of the amine probe to the sample heated to 450 °C also generates the appearance of the 2830 cm^{-1} signal due to opening of siloxane bridges, however, the signal at 2800 cm^{-1} remains absent. This observation is in line with our hypothesis that monopodal species are absent in the samples heated to 450 °C.

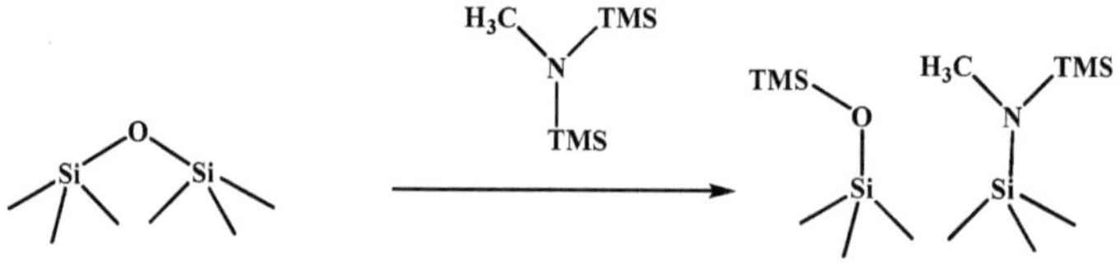

Scheme 2. Reaction of the (TMS)$_2$N-CH$_3$ probe and a siloxane bridge[286].

Additionally spectra (b) and (d) in Figure 12 on the left side show a small feature at 3450 cm^{-1} which come from a N-H stretch of the {(TMS)$_2$N(CH$_3$)H}$^+$Cl$^-$ salt, formed upon reaction of *in situ* formed HCl with excess probe molecules. The (TMS)$_2$NCH$_3$ contains 10-50 ppm water after purification which explains the *in situ* HCl formation *via* the hydrolysis of Ti-Cl bonds. In NMR, the CH$_3$ protons of this salt appear around 2.5 ppm and the ^{13}C-signal at around 24 ppm (see Figure 11). When the sample is heated to 230 °C overnight in high vacuum, this signal disappears (*i.e.* spectrum (e) in Figure 12), as the salt is fairly volatile.

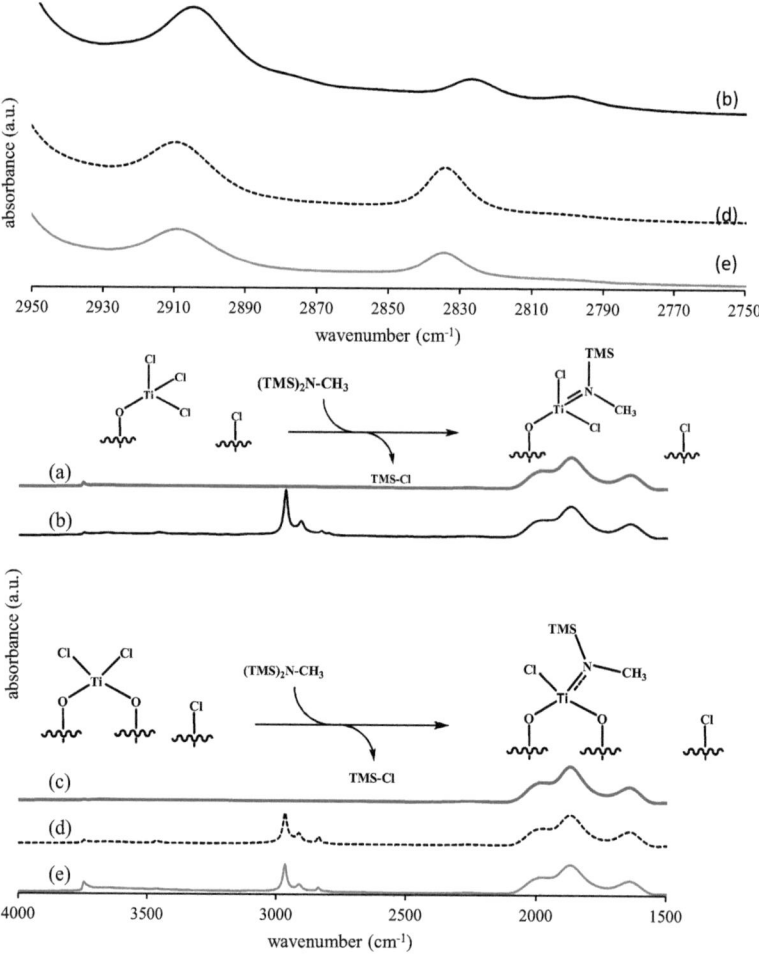

Figure 12. left: IR spectra of the TiCl$_4$-grafted materials heated to 250 °C, before (a) and after (b) grafting of (TMS)$_2$N-CH$_3$, and of the material heated to 450 °C, before (c) and after (d) probe grafting. Spectrum (e) corresponds to the same sample as spectrum (d) but after heating at 230 °C overnight. Right: Magnification of grafted spectra (b), (d) and (e).

DRS UV-Vis, the ^{35}Cl-NMR and the amine probe experiments are consistent and indicate a gradual restructuring of the Ti-species from predominantly monopodal to tripodal upon heating to 450°C.

Extended X-ray Absorption Fourier-transformed Spectroscopy (EXAFS) reveals two major scatterers for the material heated to 450 °C, attributed to Cl- and O-ligands (Figure 13, spectrum a). Comparing the EXAFS of fine anatase particles, previously reported by Fraile et al.,[287] to spectrum (a) in Figure 13 clearly disproves their presence in our material as a major phase. Comparing the first coordination shell of Ti to that of TS-1 suggests that the Ti is surrounded by more than 2 but less than 4 oxygen atoms, in line with our proposed tripodal site. We emphasize that no indication for a Ti-Ti pathway could be fitted in a satisfying manner[287]. A reasonable fit was obtained for the material heated to 450°C with approximately 3 O-atoms (at 1.8 Å) and 1 Cl-atom (at 2.2 Å) in the first shell. The second shell could not be fitted without exceeding the numbers of independent parameters. The pre-edge shows the expected feature for tetrahedral coordinated Ti-sites (electronic transition: 1s → 3pd) for the materials heated to 450°C and TS-1 (figure 13; right side: graph (a) and (b)).

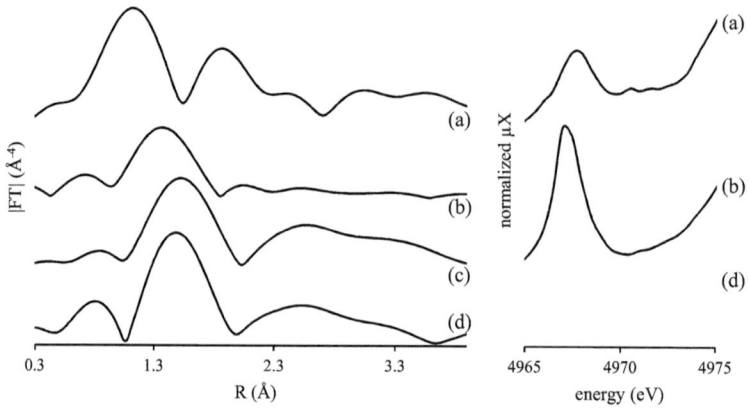

Figure 13. k^3-weighted R-space EXAFS spectra (left) and XANES spectra (right) of the TiCl$_4$-grafted material, heated to 450 °C (a), TS-1 (b), the grafted material after washing with *tert*-butanol (c) and after a catalytic experiment (d).

Mechanistic proposal

Having ruled out the formation of a Ti-dimer species, we are left with the question of how the TiCl$_4$-elimination (observed above 200 °C) takes place. Grafting CrO$_2$Cl$_2$ to silica pre-treated at 800 °C, Scott *et al.* proposed the formation of the six-membered ring-structure shown in Scheme 3, *via* transfer of a Cl-ligand from the Cr to the Si[288]. As there is no evidence for the significant abundance of the required strained hydroxyl-bearing siloxane rings at silica heated to 700 °C, the Cl-transfer hypothesis deserves further investigation.

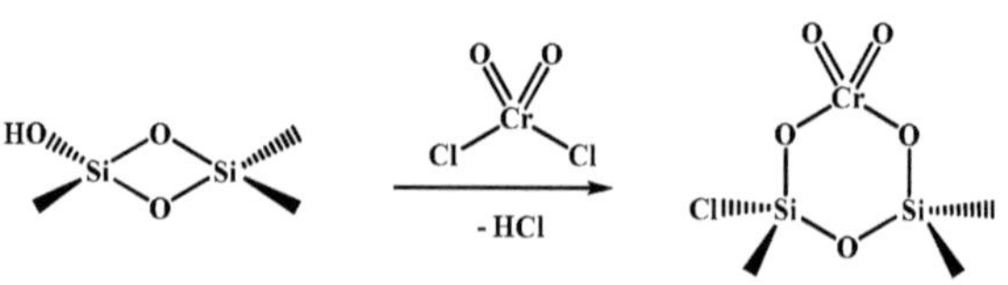

Scheme 3. Reaction of CrO$_2$Cl$_2$ with a hydroxyl-bearing strained siloxane ring[272].

Starting from the unstrained monopodal ≡SiO-TiCl$_3$, a transition state could be located for the Cl-transfer to the geminal Si-atom (see Scheme 4). However, such a rearrangement faces a thermally inaccessible barrier of 63 kcal mol^{-1}, mainly due to ring-strain.

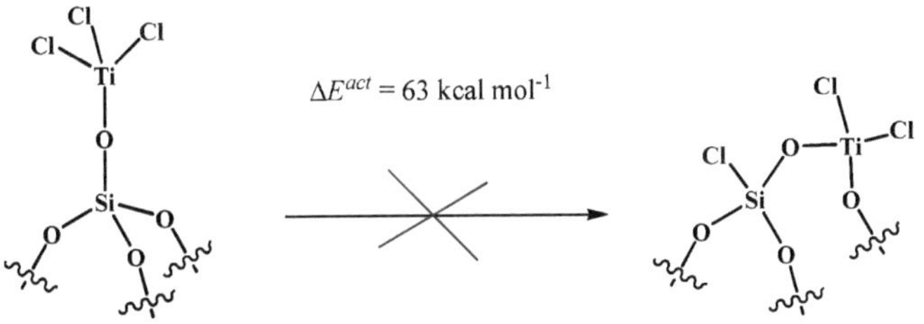

Scheme 4. Transfer of a Cl-ligand from Ti to Si for a small model.

Extending the model to a slightly bigger system (Scheme 5) helps to release the ring-strain, bringing the barrier for Cl-transfer down to 38.5 kcal · mol^{-1}. Assuming a pre-exponential factor of 10^{14} s^{-1}, the half-lifetime of this rearrangement reaction can be estimated to be ≈100 minutes at 200 °C. As a consequence of this reaction, the Cl-atom is transferred approximately 0.5 nm away from the original TiCl$_3$ group, and a fairly unstrained bipodal (≡SiO)$_2$TiCl$_2$-site is created. This mechanism is consistent with the appearance of an additional compound with high QCC in the ^{35}Cl-NMR spectrum upon heating (middle trace of Figure 7).

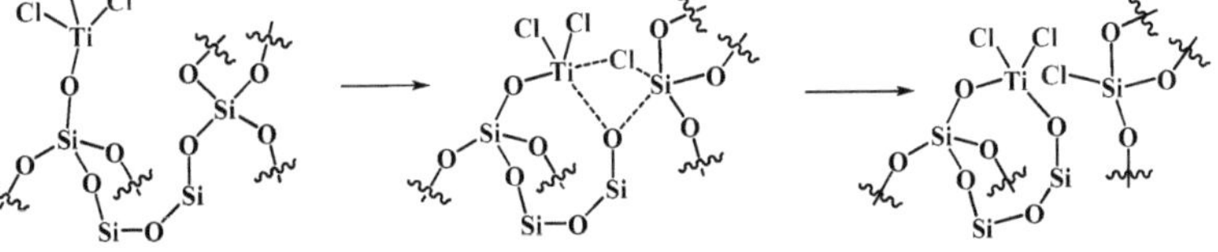

Scheme 5. Transfer of a Cl-ligand from Ti to Si for a slightly bigger surface model (see supporting information for full model).

It is our hypothesis that this shifted Cl-ligand can be picked up by a neighboring ≡SiOTiCl$_3$ site, thus forming TiCl$_4$ and re-establishing a siloxane bridge (see Scheme 6). The adiabatic barrier for this reaction is predicted to be 36.2 kcal · mol^{-1}, *i.e.* 2 kcal · mol^{-1} lower than the Cl-transfer step. This mechanism explains the elimination of TiCl$_4$, and hence the decrease in Cl/Ti ratio at higher temperatures, without the direct interaction of two ≡SiOTiCl$_3$ sites. Consecutive transfer of a second Cl-ligand to the silica surface would transform the bipodal (≡SiO)$_2$TiCl$_2$ species into tripodal (≡SiO)$_3$TiCl-sites.

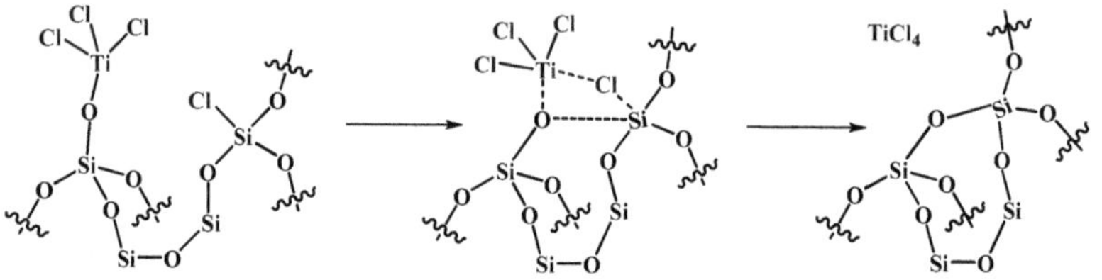

Scheme 6. Mechanism for the elimination of TiCl$_4$.

Detection of Si-Cl bonds proved to be a difficult task. Indeed, the signal-to-noise ratio of ^{29}Si MAS NMR spectroscopy is not sensitive enough to observe those species. Model compounds as (triphenylsilyl)chloride proved that Si-Cl bonds cannot be observed by ^{35}Cl NMR. A possible reason could be that the quadrupolar coupling constant for Si-Cl groups is too low leaving not enough time for refocusing in between the adiabatic WURST pulses.

The use X-ray Photoelectron Spectroscopy (XPS) in earlier work has shown that the Cl 2p orbital energies are slightly shifted from 198.4 eV for Ti-Cl species[289], and to 199.5 eV for Si-Cl species[290]. Indeed, the spectra obtained for the TiCl$_4$-grafted materials (Figure 14) heated at higher temperatures clearly reveal a broadening of the signal to higher binding energy, associated with a contribution of a second transition. This observation can be explained by the formation of an additional Cl-species with different binding energy. For the material heated to 50 °C, only the Cl-Ti signal at 198.4 eV was necessary to fit the experimental data (see figure 14). Due to spin-orbit-coupling, the signal of the 2p orbital of Cl is split into two components, $2p_{3/2}$ (red line in figure 14) and $2p_{1/2}$ (orange line in figure 14). No evidence for Si-Cl bonds could be obtained for this sample. However, the spectra of the materials heated to 450 °C could not be fitted with a single signal at 198.4 eV. Indeed, for those samples, a second signal centered at 199.5 eV becomes important, accounting for ≈50 % of the Cl 2p signal, respectively. These results are in line with our Cl-tranfer hypothesis.

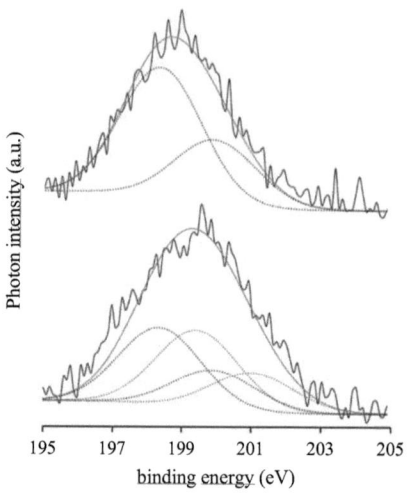

Figure 14. XPS spectra of the TiCl$_4$-grafted material after heating to 50°C (top) and 450°C (bottom); the grey lines are the simulated spectra, based on two signals each two components (2p$_{3/2}$ and 2p$_{1/2}$), positioned at 198.4 (red and orange line) and 199.5 eV (dark and light blue line).

Summarizing all experimental and computational experiments, the restructuring mechanism in Figure 15 can be put forward. A monopodal ≡SiOTiCl$_3$ species transfers one Cl-ligand to the silica surface, adjacent to the Ti-site by opening a ≡Si-O-Si≡ siloxane bridge leading to a non-strained bipodal (≡SiO)$_2$TiCl$_2$ species and ≡Si-Cl bonds (step 1). Upon transfer of the surface-bound chlorine to a neighboring ≡SiOTiCl$_3$ species a new siloxane bridge is formed, while TiCl$_4$ is eliminated (step 2). In the last step, one more Cl-ligand from the bipodal Ti^{IV} site can be transferred to the surface. This last step would establish tripodal species, explaining the final Cl/Ti ratio of 2 (ICP-OES), the ratio of Ti-Cl to Si-Cl bonds of ≈1 (XPS), the signal decrease in ^{35}Cl NMR upon heating from 250 to 450°C and the EXAFS first-shell-analysis giving a coordination number significantly smaller than 2 for Cl.

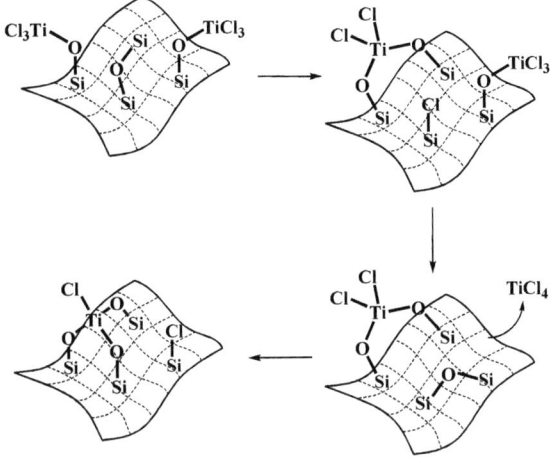

Figure 15. Proposed thermal restructuring mechanism leading to tripodal Ti^{IV}-species.

Catalytic epoxidation experiments

Correlating structure and activities in heterogeneous catalytic processes demands that one can observe differences in the kinetic activity of the TiCl$_4$-grafted samples at different temperatures. The catalytic performances of the three different materials were tested, being TiCl$_4$-grafted materials post-treated at 200 and 450 °C and a catalyst prepared *via* impregnation with Ti(O^{i}Pr)$_4$ according to a benchmarked literature procedure[291], in the solvent-free epoxidation of cyclooctene with *tert*-butylhydroperoxide as a model reaction. Prior to exposing the catalyst to the reaction mixture, the material was washed with a *tert*-butyl hydroperoxide mixture to substitute the Cl-ligand of the surface species by a *tert*-butylperoxo ligand in order to avoid any interference of chlorine. Ligand exchange can be observed with Raman spectroscopy (Figure 16), showing a disappearance of the Ti-Cl stretching vibration at 487 cm^{-1} and a slight perturbation of the symmetric deformation of the Ti^{IV}-tetrahedron from 530 to 517 cm^{-1}. The disappearance of a Cl-ligand is also in line with bulk analysis (ICP-OES) and EXAFS spectroscopy (Figure 13, left side, spectrum c). The first coordination-shell of the washed material increases to over 4. This can also be seen from the disappearance of the pre-edge feature (see figure 13, right side, spectrum c).

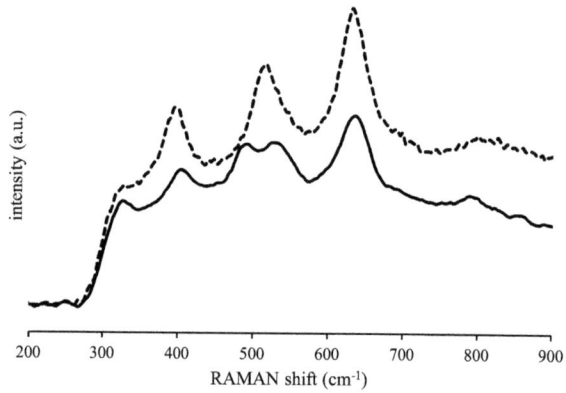

Figure 16. Raman spectrum of the material heated to 450 °C, mainly containing (≡SiO)$_3$TiCl, before and after washing with *tert*-butanol ((≡SiO)$_3$TiOR).

The epoxidation activity of the materials was compared by placing a catalyst amount into the catalyst bed which corresponds to 0.3 mg of Ti, due to the different Ti-loading of the various materials. The contact time of 2 minutes was chosen in such a way that the reactor effluent still contained 2

% of the limiting reactant (*i.e.* the peroxide) when the most active catalyst (*i.e.*, the $TiCl_4$-grafted material heated to 450 °C) was used. The conversion of the three catalysts during 20 days on stream are shown in figure 17. The $TiCl_4$-grafted material which was heated to 450 °C is significantly more active than the material heated to only 250 °C. Furthermore it can be observed that the latter material strongly deactivates. After two weeks on-stream, no significant activity was observed. ICP-OES analysis reveals a Ti loss of ±95 % after two weeks on stream. The material prepared by impregnation with $Ti(O^iPr)_4$ showed an intermediate activity, but also suffers from deactivation. This is probably due to leaching since ±30 % less Ti was observed by ICP after two weeks on-stream. The catalytic data in Figure 16 demonstrates unambiguously that regardless of the speciation of the impregnated catalyst, the material is not as stable as the $TiCl_4$-grafted material, heated to 450 °C for which ample evidence was provided that the active sites are multipodal. No changes between the butanol washed T450 and T450 used for epoxidation could be observed in XANES or EXAFS (see Figure 13, spectra (c) and (d)), indicating that the material heated to 450°C is stable under the defined reaction conditions.

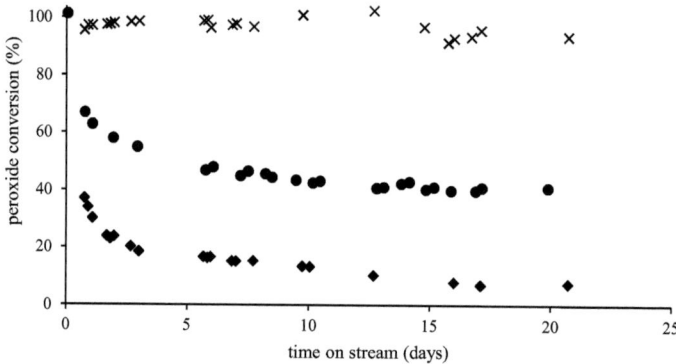

Figure 17. Stability of various catalysts during the solvent-free epoxidation of cyclooctene with *tert*-butyl hydroperoxide (100 mM) at 80 °C: TiCl$_4$-grafted materials heated to 450 °C (×), and 250 °C (♦), and a Ti(O^{i}Pr)$_4$-impregnated catalyst (●).

2.3 Conclusions

Grafting titanium tetrachloride to isolated silanol groups on dehydrated silica was investigated with a variety of techniques. Initially, site-isolated ≡SiOTiCl$_3$ species are formed, which can rotate around the Si-O-Ti axis, thereby giving rise to a narrow signal in the ^{35}Cl-NMR powder spectrum. Upon heating, Cl-transfer from those monopodal species to the silica surface results in the formation of multipodal species, which are no longer able to rotate. The further transfer of a Cl-ligand on the silica surface to a neighboring ≡SiOTiCl$_3$ species leads to the elimination of TiCl$_4$ and explains the decrease in Ti-loading at higher temperatures. Eventually, tripodal (≡SiO)$_3$TiCl species are formed on the silica surface by transferring a second Cl onto the silica surface. The corresponding (≡SiO)$_3$TiOOR species, obtained by washing the (≡SiO)$_3$TiCl species with ROOH, are shown to be very active and stable towards leaching in the solvent free epoxidation of cyclooctene with *tert*-butylhydroperoxide under continuous flow conditions.

3 Grafting CrO_2Cl_2 and $VOCl_3$ onto Silica

The volatile metal precursors CrO_2Cl_2 and $VOCl_3$ were grafted to silica dehydrated at 700°C in order to prepare monopodal $\equiv SiO\text{-}MO_xCl_{y-1}$ species (M=V, Cr). Restructuring of the obtained surface complexes were investigated up to 450 °C with different spectroscopic techniques (viz., NMR, UV-Vis, IR, Raman). This thermal restructuring involves partial elimination of $VOCl_3$ or CrO_2Cl_2. The remaining surface species become multiply bound to the silica surface balancing the lost Chlorine ligand transferred to an eliminated metal chloride.

3.1 Introduction

Chromium- and vanadium-containing solid catalysts are widely used for industrial reactions, amongst other important transformations[297-305]. Because the nature of the surface species in such heterogeneous catalysts is often ill defined, they are an important subject of scientific disucssions. In fact, diverse surface morphology often leads to a multitude of surface species, each of which may exhibits different levels of activity and/or selectivity. This is of particular relevance as synthesised catalysts are often subjected to heat treatment procedures, in order to increase activity and/or the stability of the catalyst. In principle, high catalytic activity is related to the 'site-isolation', *i.e.* in order to obtain the highest levels of activity, each metal centre should be isolated thus ensuring the highest surface area and accessibility of the active metal species[275,159,276]. A popular approach towards the synthesis of such well-defined catalysts is grafting of a volatile metal precursor, such as CrO_2Cl_2 and $VOCl_3$, to the isolated silanol ($\equiv$SiOH) groups of thermally pre-treated silica. The reaction between the volatile metal chloride and the $\equiv$SiOH species leads to both the elimination of HCl, and the formation of isolated, monopodal *i.e.* singly bound, $\equiv$SiO-MO$_x$Cl$_{y-1}$ species (M=V, Cr). The aim of this contribution is both to prepare such well-defined Cr(VI) and V(V) species, and to study their thermal rearrangement. This approach should in principle allow one to control the molecular-level architecture of such materials, potentially yield uniform active sites for studying catalytic transformations, and thereby allow more relevant structure-activity relationships to be developed[306,307].

3.2 Results and Discussion

3.2.1 Grafting of CrO_2Cl_2

Figure 18 shows the nearly complete consumption of the isolated $\equiv$SiOH signal at 3745 cm^{-1} upon grafting of CrO_2Cl_2 at room temperature. Analysis by ICP OES describes a Cr-loading of 1.43±0.05 wt%, or 0.8 Cr-species per nm^2. This surface loading is in good agreement with the density of isolated silanol sites on silica dehydrated at 700°C, and confirms that the CrO_2Cl_2 stoichiometrically reacts with those $\equiv$SiOH groups[277]. HCl gas (reaction 1) was formed *in situ* and identified by gas phase IR spectroscopy via its characteristic rotational-vibrational manifold at 2900 cm^{-1}.

$$\equiv SiOH + CrO_2Cl_2 \rightarrow \equiv SiOCrO_2Cl + HCl \quad (1)$$

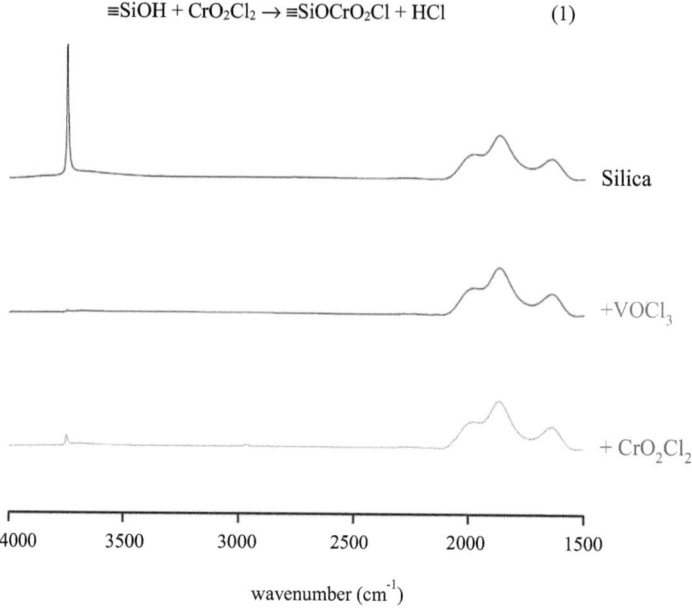

Figure 18. Transmission IR spectra of silica dehydrated at 700 °C, before and after grafting of CrO_2Cl_2 and $VOCl_3$ at room temperature (samples are degassed at 50 °C).

The Cl/Cr ratio equals 1±0.1, in line with the stoichiometry in reaction 1. The static ^{35}Cl-NMR spectrum of the material (Figure 19) shows only one compound with a QCC of 10 MHz. Static simulation of the QCC for $\equiv$SiO-CrO$_2$Cl species shown in Figure 20 result in a value of 30 MHz[308]. Taking into account the rotation around the Cr-OSi and Si-O bonds, multiple spatial orientations of the Cl-atom are averaged out and the resulting QCC of only 11 MHz is in good agreement with the experimental value.

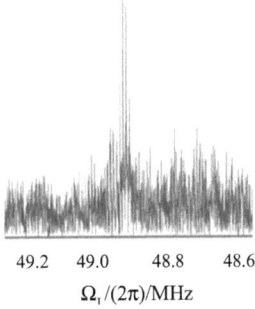

Figure 19. Static ^{35}Cl-NMR spectra of the Cr50 sample (recorded on a 500 MHz spectrometer).

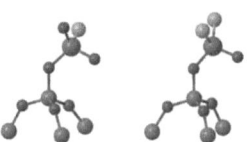

Figure 20. DFT-optimized models for ≡SiO-VOCl$_2$ and ≡SiO-CrO$_2$Cl species.

Subsequently the influence of a thermal treatment was investigated, with the aim of improving the stability of the Cr(VI) species. ICP-OES elemental analysis reveals a steady decrease in metal loading, from about 1.43 wt% to 0.75 wt% after heating the material to 450 °C, *i.e.* a loss of nearly 50 %. Simultaneously, a strong decrease in Cl/Cr-ratio from 1 to 0.3 was observed (Figure 21).

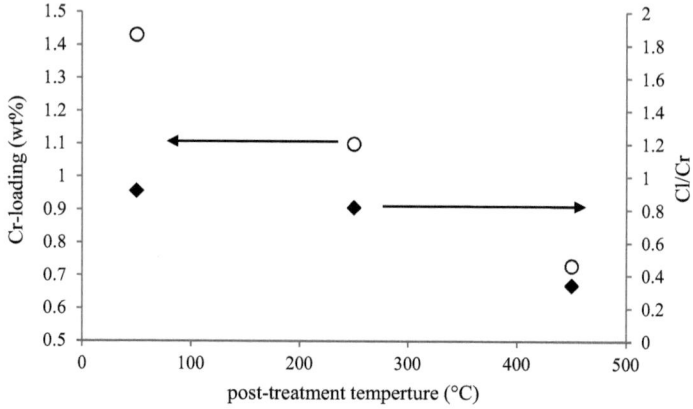

Figure 21. Decrease of the Cr-loading and the Cl/Cr ratio during thermal post-treatment.

The gas phase above the material during a TPD-experiment was monitored by transmission IR spectroscopy in a sealed and evacuated glass vessel equipped with two KBr-windows. Two distinct signals at 500 and 1000 cm^{-1} could be observed, increasing in intensity at higher post-treatment temperature (Figure 22). Those two signals are characteristic for the fundamental Cr-Cl and Cr=O stretching vibrations of chromyl chloride CrO_2Cl_2[309a]. The elimination of CrO_2Cl_2 is suggested to be the result of a chemical restructuring reaction. Physisorbed CrO_2Cl_2 can be excluded as Cl/Cr ratio was 1 for the Cr50 sample, and the UV-Vis spectra do not provide evidence for physisorbed CrO_2Cl_2 (Figure 23)[309b]. Notably, the formation of CrO_2Cl_2, in combination with a 50% decrease in overall Cr-loading, and the drastic decrease in Cl/Cr ratio, suggest that half of the ≡SiO-CrO_2Cl species transfer their Cl-ligands to the other half, which would then form CrO_2Cl_2 being released into the gas phase. Interestingly, all investigated samples are EPR silent, thus excluding the formation of Cr(V) or Cr(IV).

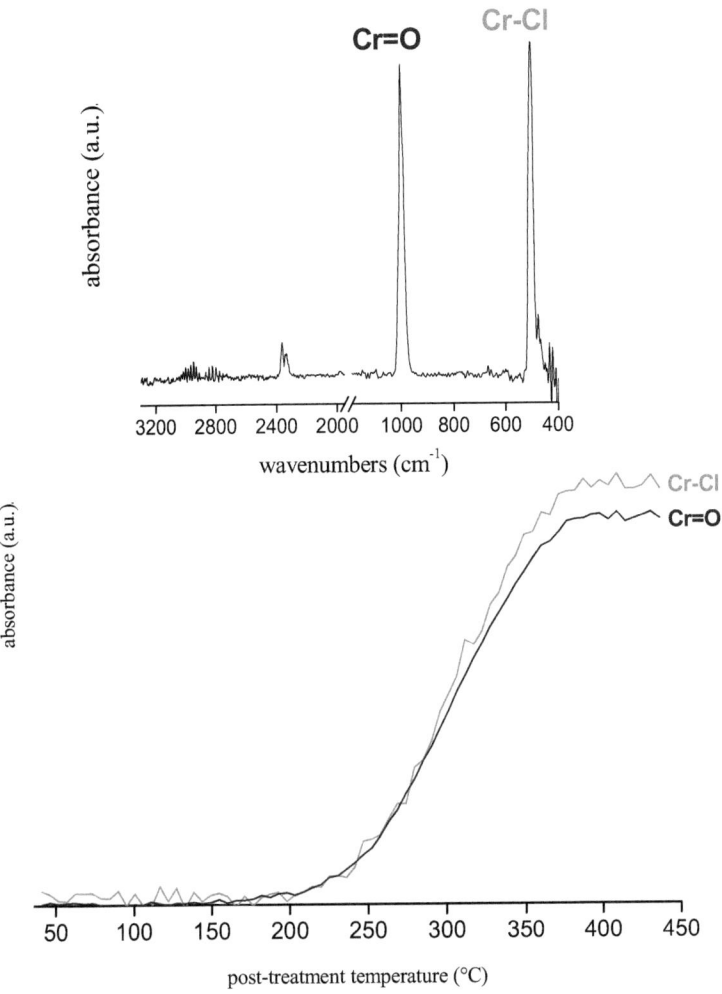

Figure 22. Formation of CrO_2Cl_2 during thermal post-treatment; increase of the Cr-Cl and the Cr=O stretch at 500 and 1000 cm^{-1}, respectively.

Raman spectroscopy reveals the (unresolved) symmetric and asymmetric Cr=O stretches of the ≡SiO-CrO$_2$Cl species in the Cr50 sample at 980 cm^{-1}, i.e. 20 cm^{-1} lower than for the free CrO$_2$Cl$_2$ reactant. The intensity of that signal increases significantly for the Cr450 sample, despite the fact that the Cr-loading reduced by 50 % (Figure 21). The reason for this is that the ν(Cr-OSi) vibration, coincidentally occurring at the same frequency, has a M06-DFT-predicted five-times larger Raman activity than the ν(Cr=O) in the monopodal species. The signals denoted with * around 800 and 1050 cm^{-1} arise from the silica support. Background-corrected Attenuated Total Reflection Infrared spectroscopy (ATR-IR) confirms that the Cr=O vibrations remain intact during the heat treatment (not shown).

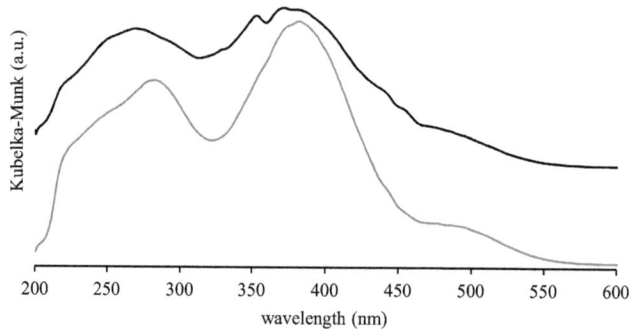

Figure 23. Diffuse Reflectance UV-Vis spectra of the Cr50 (grey) and Cr450 (black) samples.

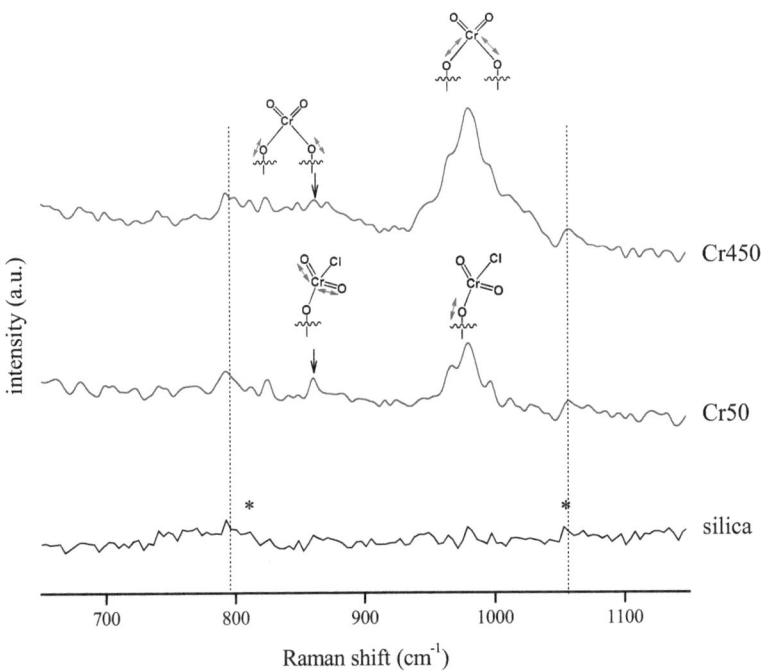

Figure 24. Raman spectra of the Cr50 and Cr450 samples.

Based on the experimental results, we suggest that an analogous mechanism as found for the ≡SiO-TiCl$_3$ system restructures the Cr-surface species. Starting with a monopodal, freely rotating ≡SiO-CrO$_2$Cl species, in line with the Cl/Cr ratio, and ^{35}Cl solid state NMR. Thermal post-treatment triggers the transfer of a Cl to the silica surface, thus leading to bipodal (≡SiO)$_2$CrO$_2$ sites. The transferred chlorine can react with a second monopodal Cr species and thereby eliminate CrO$_2$Cl$_2$ (temperature programmed IR in Figure 22), thus explaining the overall decrease in Cl by ICP-OES. No significant differences could be observed with Raman or UV-Vis spectroscopy when the Cr450 sample was kept under ambient conditions for several days, in line with a increased stability of ($\equiv$SiO)$_2$Cr(=O)$_2$ surface species.

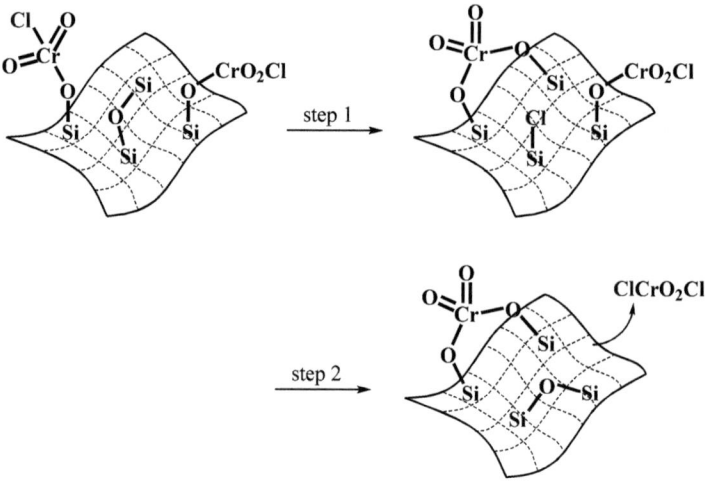

Figure 25. Thermal restructuring of isolated $\equiv$SiOCrO$_2$Cl to bipodal ($\equiv$SiO)$_2$CrO$_2$ species.

It needs to be mentioned that Scott *et al.* previously suggested the formation of bipodal chromium sites upon grafting of CrO$_2$Cl$_2$ to silica pre-treated at 800 °C[310]. The formation of a six-membered ring-structure shown in scheme 3 was suggested to occur *via* transfer of a Cl-ligand from the Cr to the Si. However, no evidence was found that the isolated silanol groups (*i.e.* the grafting sites) would preferentially be part of such highly strained siloxane rings at silica pre-treated to 700 °C as in our samples. Moreover, such a grafting mechanism would not explain the elimination of CrO$_2$Cl$_2$ from the surface, nor the decrease in Cl/Cr ratio as observed by ICP.

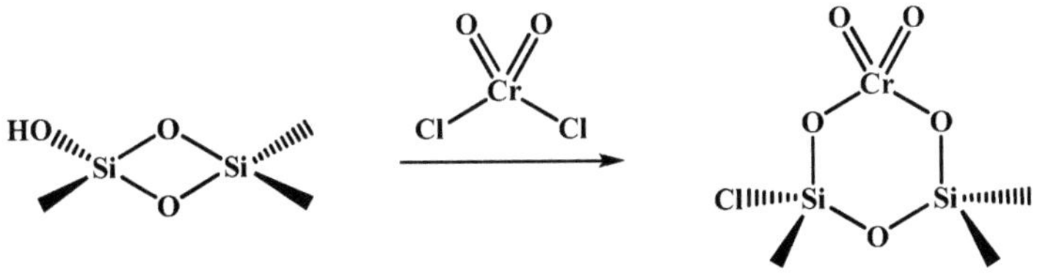

Scheme 3. Reaction of CrO_2Cl_2 with a hydroxyl-bearing strained siloxane ring.

3.2.2 Grafting of VOCl₃

Similar to CrO₂Cl₂, the grafting of VOCl₃ to pre-treated silica leads to the quantitative consumption of the isolated silanol sites (Figure 18). The V-loading of the V50 sample is 0.8 V · nm⁻² (1.4 wt%) as obtained earlier with the Cr- and Ti-loading. The ^{35}Cl-NMR spectrum in Figure 26 only shows one feature with a QCC of 3.7 MHz which is remarkably small for a solid vanadium chloride compound[288b]. Due to the rotation of monopodal vanadium surface species, the static QCC of the ≡SiO-VOCl₃ species shown in Figure 20 (22.6 MHz) reduces to only 4.1 MHz. This value agrees very well with the experimentally detected value (3.7 MHz), and suggest the formation of ≡SiO-VOCl₂ species, in line with the detected Cl/V ratio of 2±0.1.

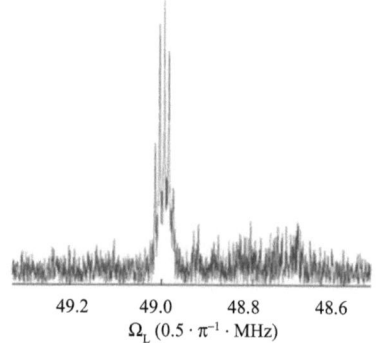

Figure 26. Static ^{35}Cl-NMR spectra of the V50 sample (recorded on a 500 MHz spectrometer).

Upon heating to 250 and 450 °C, the V-surface content decreases from 1.4 to 0.8 wt%, along with a decrease of the Cl/V (Figure 27). In combination with the disappearance of the ^{35}Cl-NMR signal, these observations suggest that the freely rotating ≡SiOVOCl$_2$ species restructures analogously to the Ti and Cr materials.

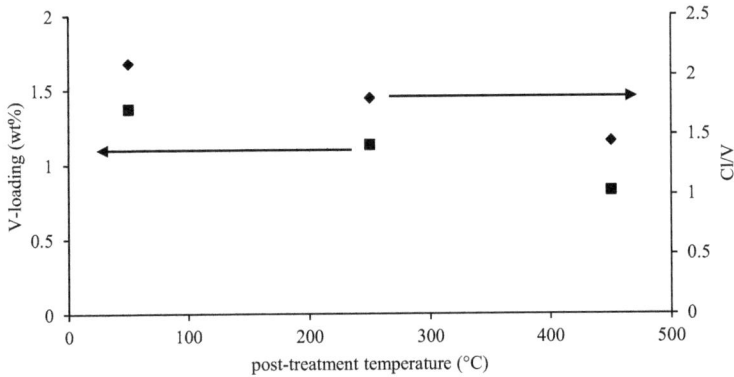

Figure 27. Decrease of the V-loading (■) and the Cl/V (♦) ratio during thermal post-treatment.

In line with this hypothesis the elimination of VOCl$_3$ was observed by gas phase transmission IR spectroscopy when the material is heated above 200 °C in a TPD experiment (Figure 28); the characteristic VCl$_3$ rocking vibration at 509 cm^{-1}, and the V=O stretch at 1043 cm^{-1} are clearly be seen (inlet figure 28)[311].

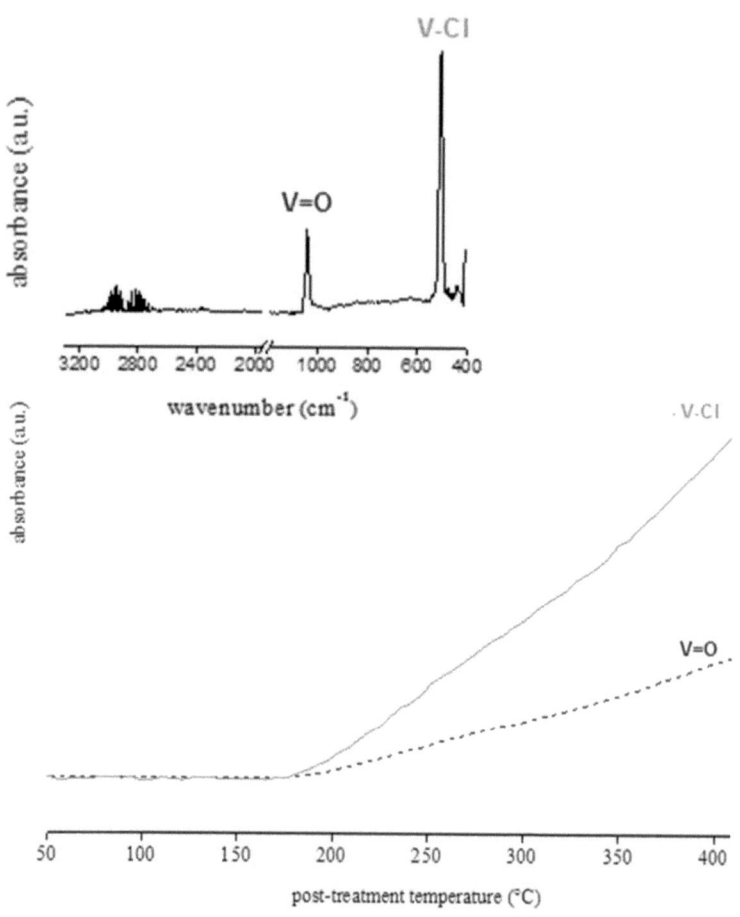

Figure 28. Formation of VOCl$_3$ during thermal post-treatment; increase of V-Cl and V=O vibration at 509 and 1040 cm^{-1}, respectively.

UV-Vis spectroscopy (Figure 29) verifies that no physically adsorbed $VOCl_3$ is present on the silica surface (no absorbance over 400 nm), and also indicates a significant change in the surface speciation with increasing post-treatment temperature. The intensity of the spectrum decreases below 220 nm, and simultaneously increases around 290 nm upon heating. In figure 30 the spectra of three reference compounds can be seen: Na_3VO_4 (containing isolated tetrahedral V^V-species), NH_4VO_3 (containing chains of tetrahedral V^V-species with V-O-V bonds), and V_2O_5 (bulk oxide with octahedral V^V). Based on these reference spectra, the formation of oligo- or polymeric vanadium species are excluded. It is suggested that V is present as isolated vanadium sites in a tetrahedral coordination for V50, in line with the literature[312-315]. However, the observed red-shift of the maximum implies a structural change of the V-species upon thermal post-treatment.

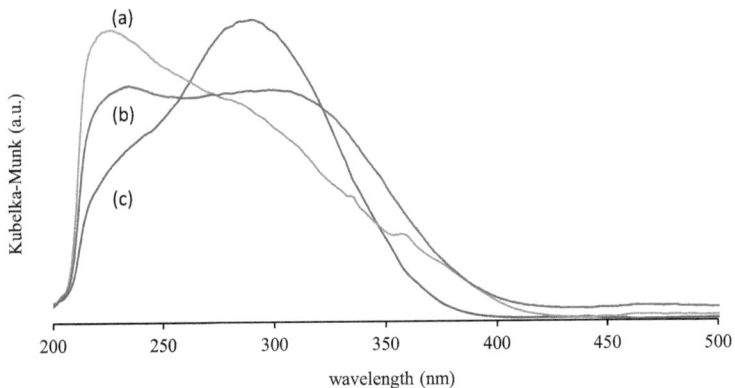

Figure 29. Diffuse Reflectance UV-Vis spectra of the V-grafted samples, post-treated at 50 (a), 250 (b) and 450°C (c).

In order to gain more insight in the V-surface speciation, the materials were exposed to a bis(trimethylsilyl)(methyl)amine probe with a ^{13}C-labeled methyl-group. During this ligand probing, the formation of TMS-Cl for the V50 and V250 sample was observed by ^{1}H NMR in the liquid nitrogen trapped effluent of the reactor. This indicates a reaction between the vanadium surface species and the probe molecule. In contrast to V50 and V250, for V450, no TMS-Cl was detected. The dominant signals in the IR spectra between 2900 and 3000 cm^{-1} (Figure 31) are due to TMS-vibrations while signals below 2900 cm^{-1} correspond to the labelled methyl group.

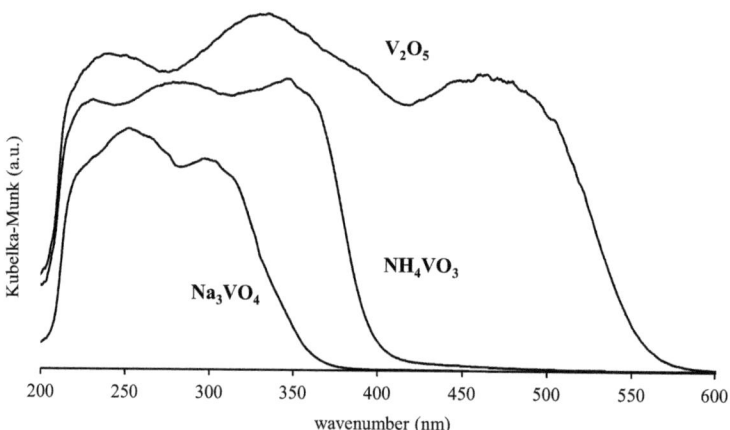

Figure 30. Diffuse Reflectance UV-Vis spectra of reference compounds.

For V450, a single signal at 2833 cm^{-1} is observed, which is attributed to a ligand reacted with a siloxane bridge (scheme 4)[316]. This assignment is confirmed by a control experiment in which thermally dehydrated silica is exposed to the amine probe and the same signal at 2833 cm^{-1} is present. For

V50, and especially V250, two other signals are observed at 2850 and 2770 cm^{-1} which are further investigated with ^{1}H MAS NMR.

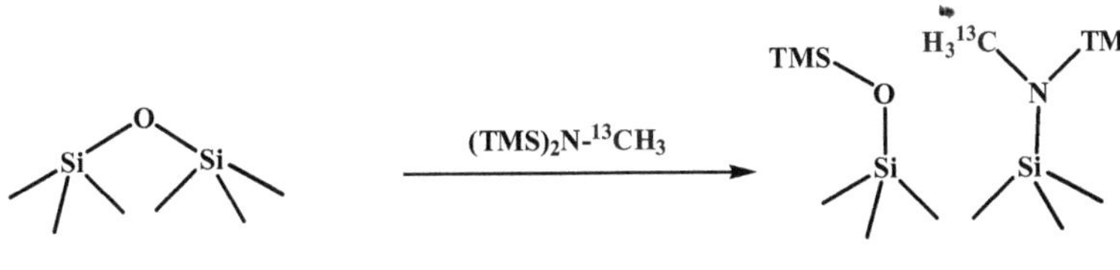

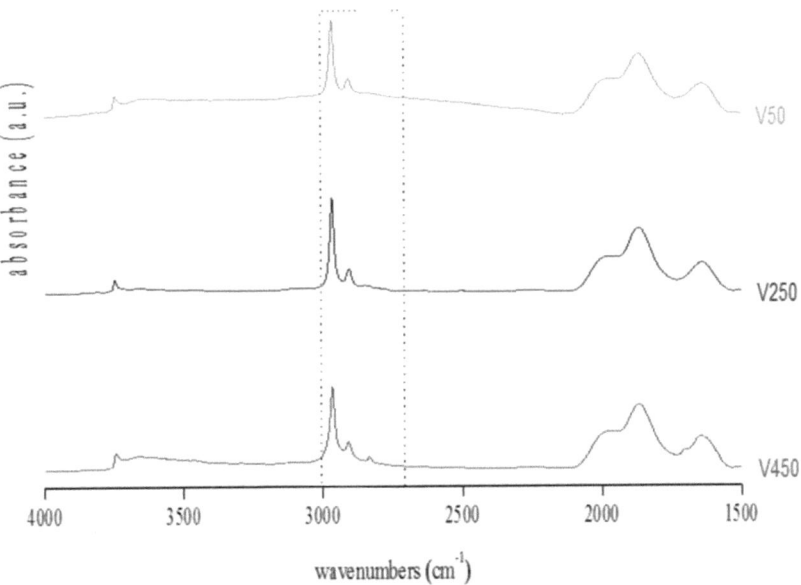

Scheme 4. Reaction of the (TMS)$_2$N-CH$_3$ probe a siloxane bride.

Figure 31. Transmission IR spectra of V50, V250 and V450 after grafting of the amine ligand.

Figure 32 displays the ^{1}H MAS NMR spectra of the three V-samples, after exposure to the (TMS)$_2$N-^{13}CH$_3$ probe. Analogous to the IR spectrum, only one significant signal can be detected for the V450 sample at around 2.5 ppm which is also present when dehydrated silica is exposed to the probe, and confirms the assignment as an ≡Si-N(TMS)-CH$_3$ group (see scheme 4)[316]. For the V50, an additional signal is observed around 4 ppm[286], assigned to the IR signal at 2850 cm^{-1}. The signal most likely corresponds to a methoxy species –OCH$_3$, as the C-H vibrations are shifted to slightly higher wavenumbers than for the amine >N-CH$_3$ group. For the V250, an additional signal is visible around 5 ppm[317]. This signal assigned to an imine species (=N-CH$_3$) due to the low corresponding IR vibration at 2770 cm^{-1}.

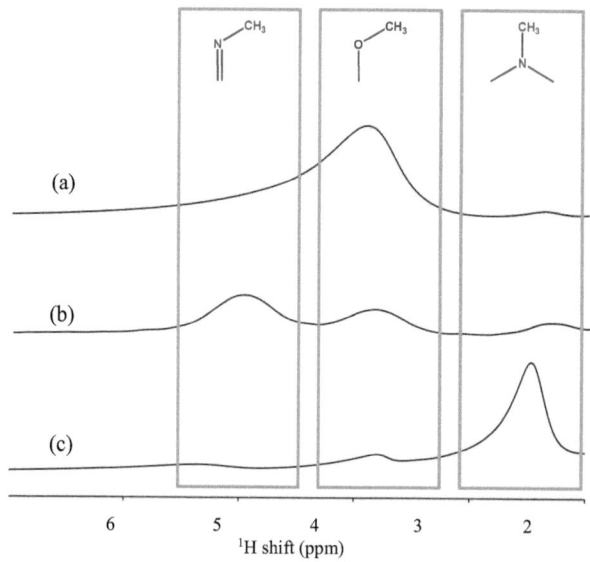

Figure 32. ^{1}H MAS NMR spectra of the V50, V250 and V450 materials after exposure to the (TMS)$_2$N-^{13}CH$_3$ probe; exp. details: Spinning rate: 12 kHz, pulse length: 4 µs, recycle delay: 2 s; NS: 64.

Figure 33 shows the 2D HETCOR MAS NMR spectrum of the V250 sample after exposure to (TMS)$_2$NCH$_3$. It can be seen that the signal at 2.5 ppm actually corresponds to two different species, *i.e.* not only ≡Si-N(TMS)CH$_3$, but also the hydrochloride salt {(TMS)$_2$N(H)CH$_3$}$^+$Cl$^-$.

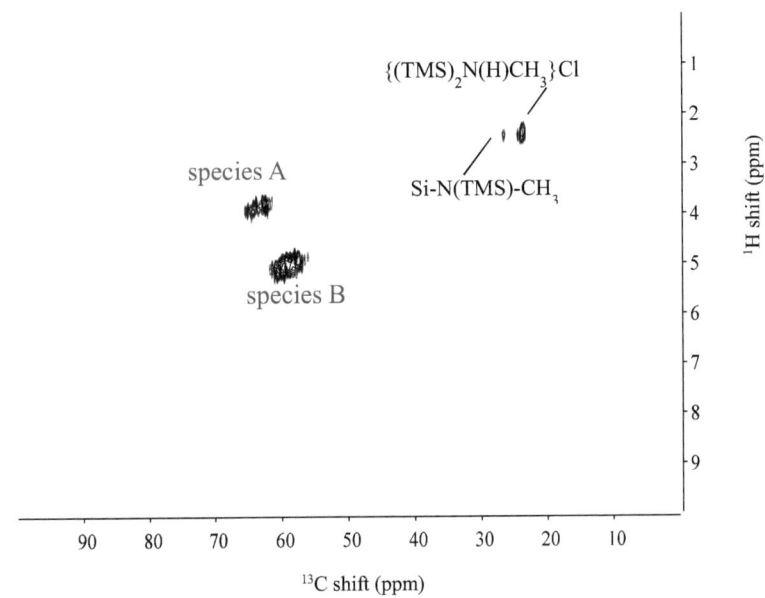

Figure 33. 2D HETCOR ^{1}H^{13}C CP MAS NMR spectrum of the V250 sample, after exposure to the (TMS)$_2$N-^{13}CH$_3$ probe; exp. details: Spinning rate: 12 kHz, pulse length: 3.1 μs, recycle delay: 2s, NS: 4096.

Based on DFT-predications (M06/6-31(d,p)-level), two molecular structures are suggested shown in Figure scheme 5 (X=Cl or - OSi$_{surface}$). Unfortunately, it is not possible to distinguish between mono- and bipodal species by ^{1}H MAS NMR. Nevertheless, neither of the two signals is present for the V450 sample, indicating that the amine probe is not reacting with the V-species present in that sample. The assignment to methoxy and imine species explains why the protons at 4 ppm correlate with the carbon signal at 63 ppm, and the protons at 5 ppm correlate with the carbon signal at 58 ppm.

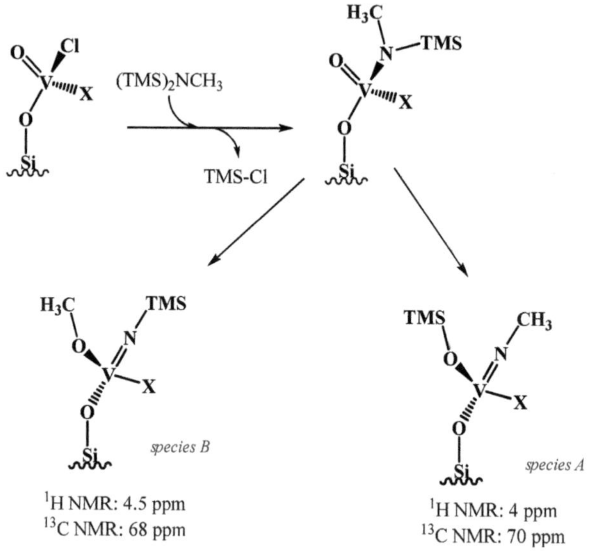

species B
^{1}H NMR: 4.5 ppm
^{13}C NMR: 68 ppm

species A
^{1}H NMR: 4 ppm
^{13}C NMR: 70 ppm

Scheme 5. Reaction of V surface sites with the (TMS)$_2$NCH$_3$ probe and the corresponding DFT-predicted NMR shifts of the methyl group protons and carbons (X=Cl or -OSi$_{surface}$).

The V-containing samples have further been analyzed with Raman spectroscopy. It can be observed that a gradual and complete disappearance of the v(V=O) at 1027 cm^{-1} (slightly shifted from 1043 cm^{-1} in VOCl$_3$) and its first overtone 2v(V=O) at 2042 cm^{-1} (Figure 34) is observed after heating, in line with previous observations in IR spectroscopy on V/CeO$_2$[318,319].

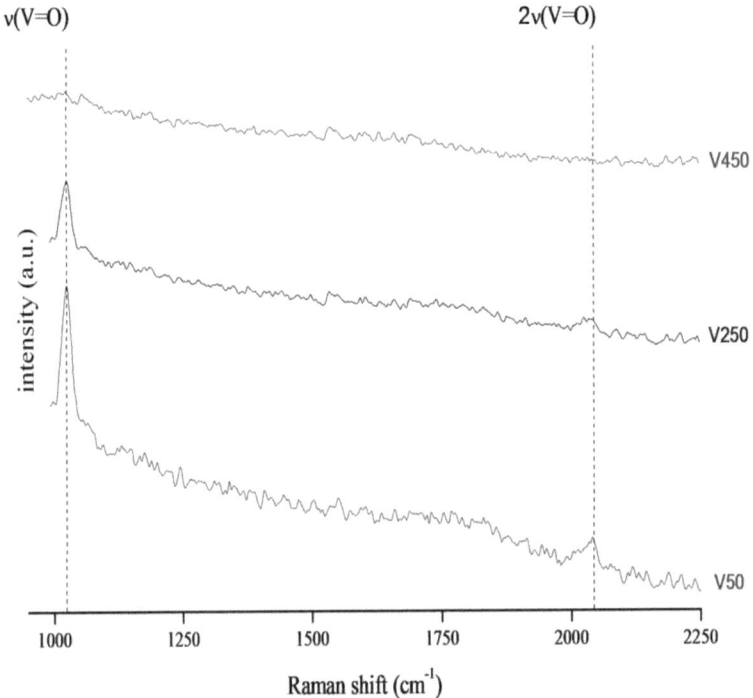

Figure 34. Raman spectra of the V50, V250 and V450 samples.

The ^{51}V MAS NMR of the V50 sample (see Figure 35) shows a true peak maximum at -292 ppm with spinning side bands (positions are shifting when the spin rate is changed). The resonance at -292 ppm is assigned to the ≡SiO-

VOCl$_2$ species, the chemical shift of which is shifted compared to the VOCl$_3$ reference. This assignment is in line with the work of Scott *et al.*[306], and further confirms that there is no VOCl$_3$ physisorbed on the V50 sample. (≡SiO)$_2$VOCl species, predicted to have a chemical shift of ca. - 500 ppm[306], can also be excluded (see fast decay of the side spinning bands in figure 35 indicating a small quadrupole moment). Remarkably, the intensity of this -292 ppm signal decreases during the thermal post-treatment. It is nearly absent in the V450 sample and a broader signal appears underneath it which is indicating a non-rotating V-species. The decrease of the signal at -292 ppm can be correlated with the decrease in ν(V=O) stretching vibration observed with Raman (Figure 36). While Raman spectroscopy is solely monitoring the V=O stretch, the ^{51}V NMR signal is ascribed to the rotating V surface species. A correlation of both signals, shows a simultaneous decrease of the molecular rotation and the V=O vibration at higher post-treatment temperature. This strongly indicates that the V450 sample contains fundamentally different species than the V50 and V250 samples.

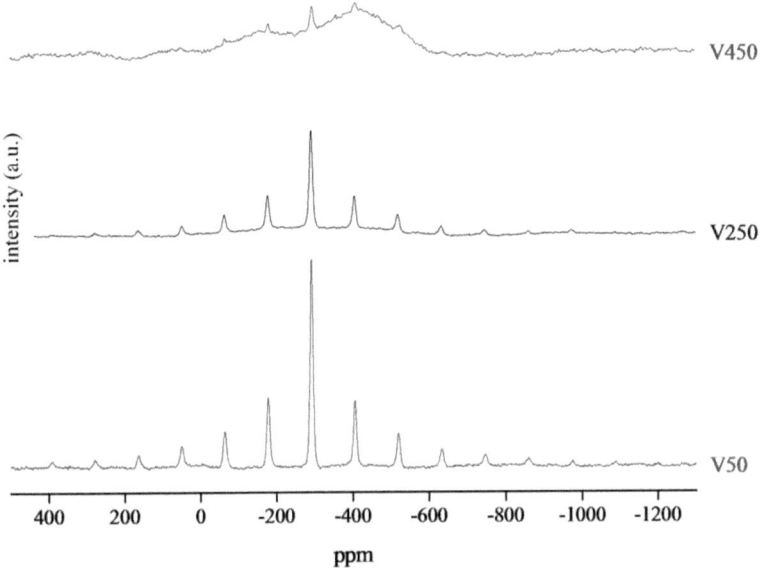

Figure 35. ^{51}V MAS NMR of the V-grafted samples.

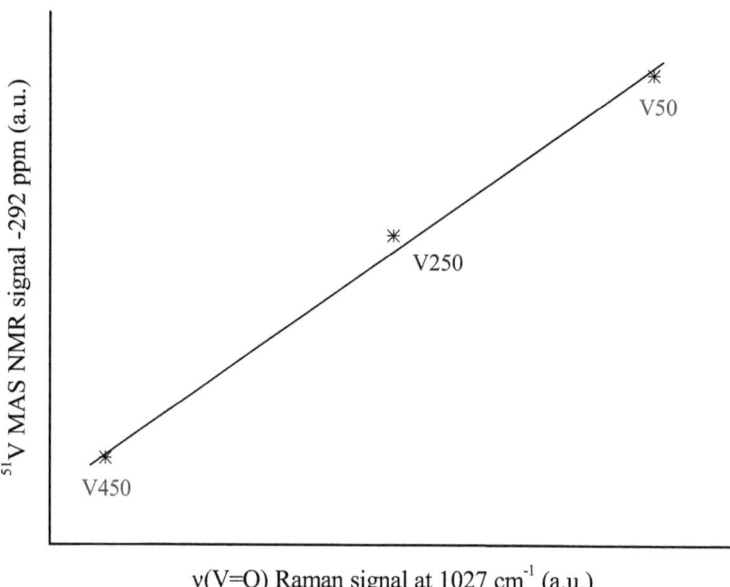

Figure 36. Correlation between the [51]V MAS NMR signal at -292 ppm (Figure 35) and the ν(V=O) stretch observed with Raman (Figure 34).

In X-ray Photoelectron Spectroscopy (XPS) the peak maxima shifts to higher binding energy for V50 (Figure 37) and V450 (Figure 38) materials, which is associated with a significant broadening of the signal. A plausible explanation for this is the formation of an additional Cl-species with different binding energy. Two different chlorine species at about 199.7 and 201.3 eV, respectively, were needed to fit the corresponding spectra. For V50, one compound is almost sufficient for an acceptable fit. However for the fit of V450 two equally important Cl-species are necessary. As Si-bound chlorine is expected at higher binding energies[320], this XPS data suggest comparable amounts of V-Cl and Si-Cl (about 2:1) species for the V450 material. The high amount of Si-Cl species indicates that not every Cl on the silica surface further reacts with another monopodal ≡SiO-VOCl$_2$, and thereby eliminates VOCl$_3$. This observation is also confirmed by ICP where only 40% of the vanadium disappears upon thermal post-treatment.

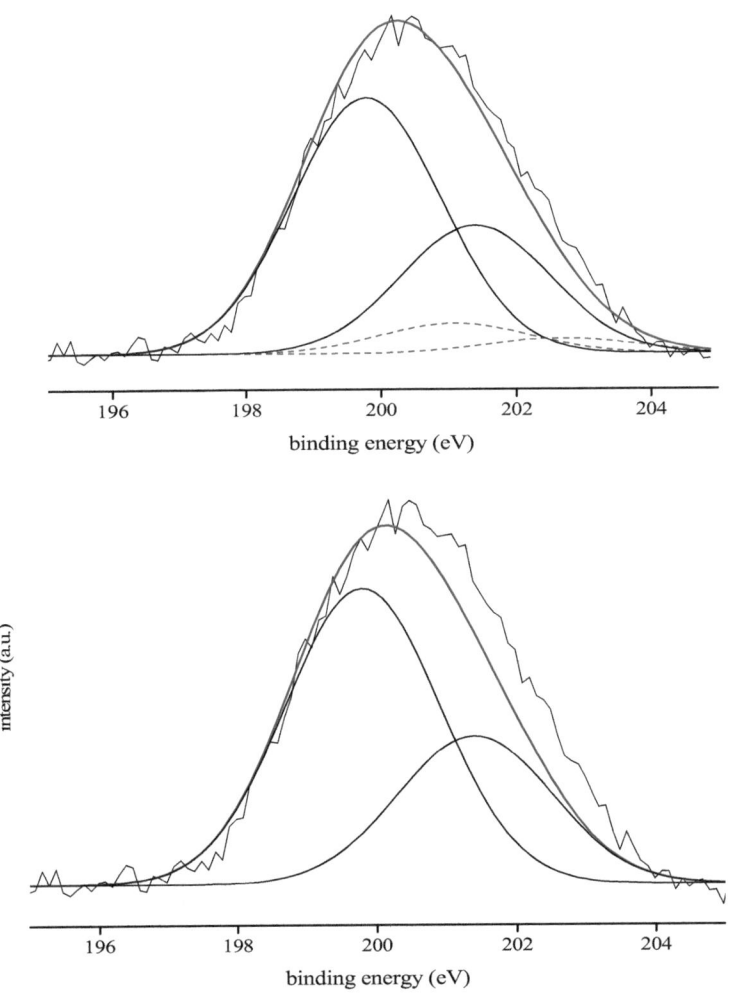

Figure 37. XPS spectrum of the V50 sample, deconvoluted with one (bottom) and two (toop) compounds.

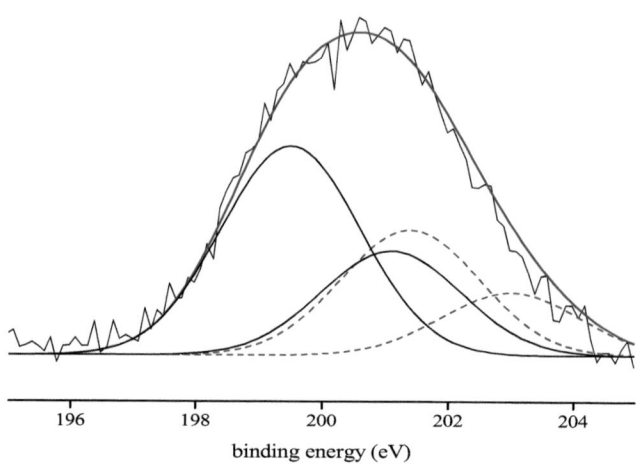

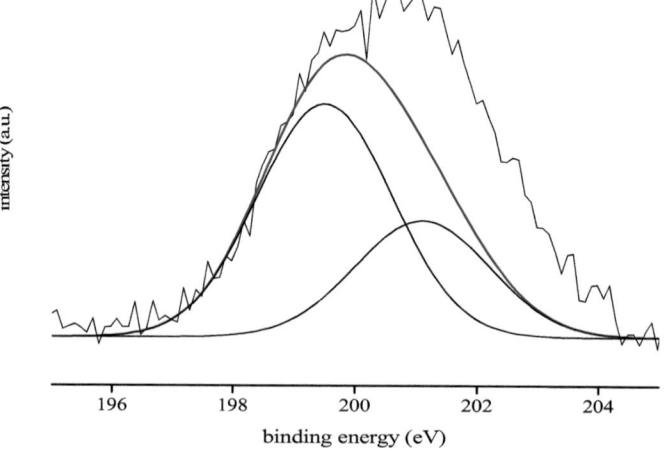

Figure 38. XPS spectrum of the V450 sample, deconvoluted with one (bottom) and two (top) compounds.

Analyzing all analyzed data, it is proposed that the ≡SiO-VOCl$_2$ species is able to react in two different manners. The first reaction pathway would be the transfer of a Cl-ligand from the V-species to the silica surface, generating Si-Cl bonds. The combination of such Si-Cl bond with another ≡SiO-VOCl$_2$ species leads to the elimination of VOCl$_3$, as observed with IR spectroscopy (Figure 28). This mechanism, shown in Figure 39, is similar to the mechanism proposed for the rearrangement of ≡SiO-TiCl$_3$ and ≡SiO-CrO$_2$Cl. However, an alternative pathway for the ≡SiO-VOCl$_2$ species is the reaction of the V=O bond with a siloxane bridge, resulting in (≡SiO)$_3$VCl$_2$ species (Figure 40). When V450 (white colour) is exposed to water, no change in UV-Vis is observed while V50 experiences a red shift and turns immediately yellow. This indicates that V450 has a more rigid coordination environment and is chemically more stable than V50, in line with the formation of multiple bonds to the silica surface upon thermal post-treatment.

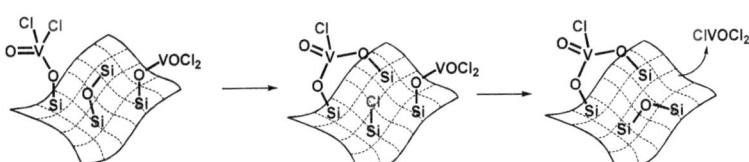

Figure 39. Transfer of a Cl-ligand from the vanadium site to the silica surface and elimination of VOCl$_3$.

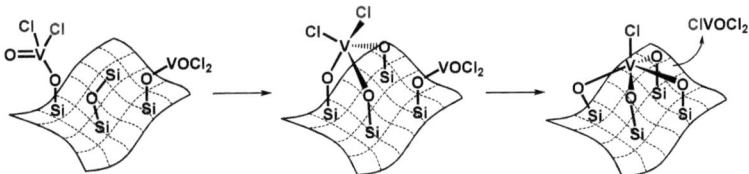

Figure 40. Reaction of the V=O bond of the vanadium site with a surface siloxane bridge.

Although no direct observation indicates which mechanism is faster, the linear correlation between the decrease of the V=O stretch and the decrease in freely rotating ≡SiO-VOCl$_2$ species observed with ^{51}V MAS NMR (see Figure 36) suggests a rather facile reaction of the V=O bond with the silica surface (Figure 40). This is most likely due to the fact that V(V) prefers a higher coordination number than 4, in line with literature reports[321,289]. The UV-Vis red shift is also in agreement with ligand field theory, which predicts a lowering of the xz- and yz-orbital energy of vanadium upon increased coordination from a tetrahedral to a more square-pyramidal environment. The LUMO (*i.e.*, the xz- and yz orbitals) thus is lower in energy, decreasing the energy necessary for a ligand-to-metal-charge-transfer. Subsequently, VOCl$_3$ is still able to be eliminated as observed experimentally, and to produce a pentavalent V^V-site with vanadium four-fold bound to the silica surface. Nevertheless, the transfer of Cl-ligand to the silica cannot be neglected either, since the XPS measurements suggest the formation of fair amount of Si-Cl bonds during the post-treatment.

3.3 Conclusions

Room temperature grafting of metal chlorides MO_xCl_y to thermally dehydrated amorphous silica featuring isolated silanol groups is a convenient method to synthesize site-isolated $\equiv$SiO-MO_xCl_{y-1} Lewis acid sites. Upon heating, thermal reconstruction takes place, forming multipodal species. Two different restructuring mechanisms could be proposed. The first mechanism proceeds *via* the transfer of a metal-bound Cl-ligand to the silica surface, opening a siloxane bridge. This leads to the formation of Si-bound chlorine, and bipodal metal sites. Reaction of the SiCl species with a second yet unrestructured monopodal species results in the elimination of MO_xCl_y and the restoration of a siloxane bridge. For $VOCl_3$-grafted materials, an additional restructuring mechanisms can take place *via* a reaction of the V=O bond with a siloxane bridge, thus forming a pentavalent vanadium species. The occurrence of this mechanism is supported by the correlated observations from ^{51}V MAS NMR and Raman spectroscopy. Most importantly, it is clear that in all the investigated cases *i.e.* $TiCl_4$-, $VOCl_3$- and CrO_2Cl_2- grafted silica, thermal restructuring of the $\equiv$SiO-MO_xCl_{y-1} species leads to multiple bonds to the silica surface, and hence an increased thermal and chemical stability.

4 Synthesis of a dimeric Ti-surface species

In this chapter, the synthesis of a silica-supported titanium catalyst with dimeric surface sites is reported. The material is prepared by sequencing the synthesis of isolated Ti-Cl surface sites, with an additional grafting of titanium isopropoxide in dry cyclohexane. The material was investigated by different spectroscopic techniques, e.g., IR, TPD IR, UV-Vis ^1H ^{13}C CP MAS NMR, XANES and EXAFS of the Ti k-edge. Computational methods were used to elaborate the molecular structure of the formed dimer sites. The stability in catalytic epoxidation with TBHP was investigated. The catalytic activity is discussed in the next chapter and is compared to the material thermally post-treated at 450°C.

4.1 Introduction

Catalytic epoxidations of olefins are one of the most important oxidative technologies applied in industry[36,64]. Currently, all large-scale catalytic epoxidation processes use Ti^{IV}-based heterogeneous catalysts to activate either organic hydroperoxide (*viz*, Styrene Monomer Propylene Oxide or SMPO technology)[33] or hydrogen peroxide (*viz.*, Hydrogen Peroxide Propylene Oxide or HPPO technology)[2]. The titanium can be grafted onto an amorphous support, or incorporated into a micro-porous crystalline material (*viz.*, TS-1). It is generally assumed that high activity is due to site-isolated tetrahedral Ti^{IV} species[228,233,272,326,327,328]. During hydrothermal synthesis, the precise molecular structure of the Ti-site is controlled by the crystallization process of the silica matrix. Different position in the matrix where the Ti can be incorporated and the occurrence of defect sites are able to create a variety of Ti-sites. However, grafting of well-defined molecular precursors onto amorphous silica allows one to investigate the activity of better defined active sites[269,275,329-334]. Considerable literature has been published on the synthesis of isolated Ti^{IV}-sites *via* grafting. Recently, Brutchey *et al.* grafted the dititanium precursor [(^{t}BuO)$_2$Ti{μ-O$_2$Si[OSi(O^{t}Bu)$_3$]$_2$}]$_2$ to mesoporous SBA-15[335]. Earlier, Scott *et al.* investigated the grafting of Ti(O^{i}Pr)$_4$ to neighboring silanol sites on silica, dehydroxylated at 200 °C[271a,288b]. According to the authors, the generated oxo-bridged Ti-O-Ti dimers are very active in catalytic epoxidations with TBHP.

4.2 Results and Discussion

Upon grafting of Ti(O^{i}Pr)$_4$ to silica grafted TiCl$_4$, which has been thermally post-treated at 450°C (referred to as Ti$_{mono}$), two signals can be observed in 2D ^{1}H ^{13}C HETCOR MAS NMR (Figure 41), corresponding to the methyl (^{13}C at 27 ppm; ^{1}H at 1.8 ppm) and methine (^{13}C at 78 ppm; ^{1}H at 5 ppm) C-H species of the *iso*-propoxy ligands[271]. These C-H species can also be observed with IR spectroscopy (Figure 42) by the stretching vibration of the C-H bonds. The Ti-loading increased by a 2.5-fold from 0.8 ± 0.1 wt% sample to 2.0 ± 0.2 wt% after grafting Ti(O^{i}prop)$_4$ onto T450. Chlorine could no longer be detected by ICP-OES analysis after digestion. The material prepared by grafting Ti(O^{i}prop)$_4$ onto T450 is referred to as Ti$_{dimer}$.

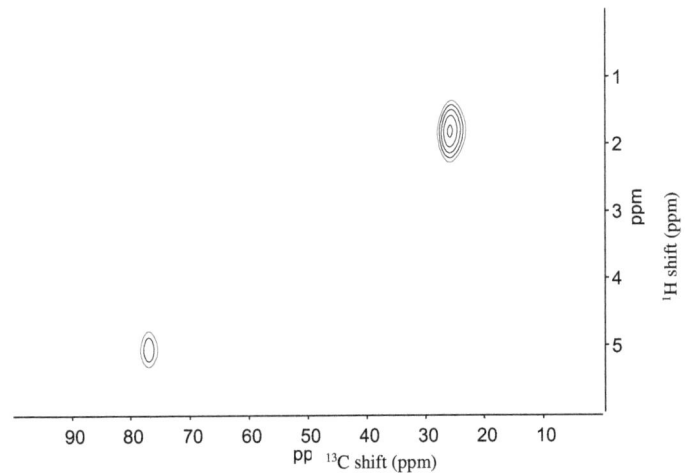

Figure 41. 2D ^{1}H ^{13}C HETCOR MAS NMR of the Ti$_{dimer}$ material.

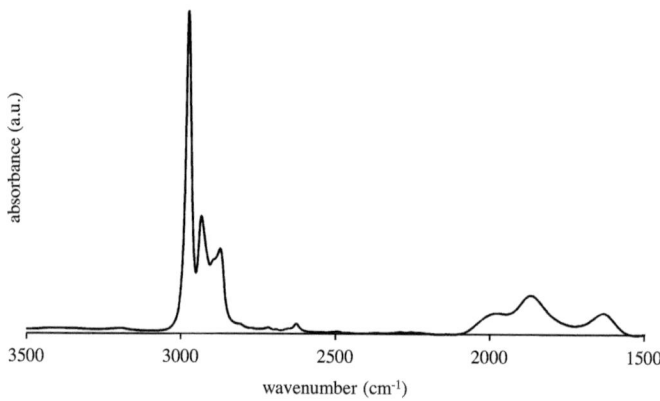

Figure 42. Transmission IR spectrum of the Ti$_{dimer}$ material showing the isopropoxy ligands.

In order to answer the question whether Ti(O^{i}Pr)$_4$ only reacts with the Ti-Cl bonds, or is also able to be inserted into (strained) siloxane bridges (*viz.*, Si-O-Si, Scheme 7)[338], Ti(O^{i}Pr)$_4$ was grafted onto silica with capped silanol groups (by heptamethysilizane). After washing with dry hexane and drying, ICP-OES revealed a loading smaller than 0.1 wt% Ti, indicating negligible reactions with siloxane bridges[286b]. Additionally, 2D ^{1}H ^{13}C HETCOR MAS NMR of the Ti$_{dimer}$ sample in Figure 41 only indicated one methane group. Indeed, the ≡Si-O^{i}Pr being formed (see Scheme 7) would appear approximately 66 ppm in ^{13}C-NMR and 4.3 ppm in ^{1}H-NMR, similar to Si(O^{i}Pr)$_4$[339]. The absence of such a signal supports our hypothesis that Ti(O^{i}Pr)$_4$ mainly reacts with the Ti-Cl bonds. Moreover, the reaction was found to be self-limiting. Using either a 50- or 2-fold excess of Ti(O^{i}Pr)$_4$, relative to the Ti-Cl bonds in the Ti$_{mono}$ material, results in the same overall Ti-loading of 2.0 ± 0.2 wt%.

Scheme 7. The reaction between Ti(O^{i}Pr)$_4$ and siloxane bridge.

When Ti(O^{i}Pr)$_4$ is grafted to silica pretreated at 700°C, the loading of the resulting material is higher than for Ti$_{dimer}$ because of a higher amount of surface anchor points (silanol groups) available on unmodified silica surfaces. Moreover, the reaction of Ti(O^{i}Pr)$_4$ with ≡SiOH produces *iso*-propanol which can open siloxane bridges[286]. In this way, new silanol groups are formed *in-situ* increasing the possible amount of grafted Ti(O^{i}Pr)$_4$ and thereby the Ti-loading.

The replacement of TiCl bonds by Ti-O-bonds is in line with the blue-shift observed in UV-Vis spectroscopy (Figure 43)[288b,328]. During the reaction of Ti(O^{i}Pr)$_4$ with Ti$_{mono}$, propene formation was observed by IR spectroscopy in the gas phase (Figure 44) at temperatures as low as 140°C. HCl is desorbed during the post-synthetic drying step in high dynamic vacuum as confirmed by ICP-OES.

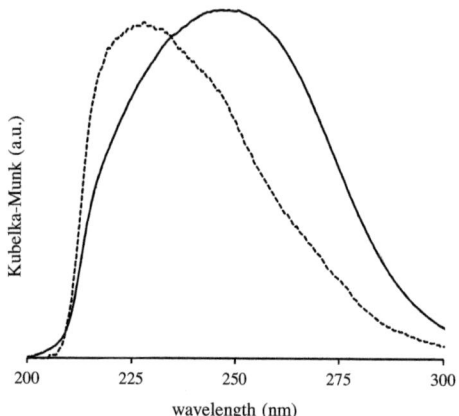

Figure 43. Diffuse Reflectance UV-Vis spectra of the Ti_{mono} (solid line) and Ti_{dimer} catalyst (dotted line).

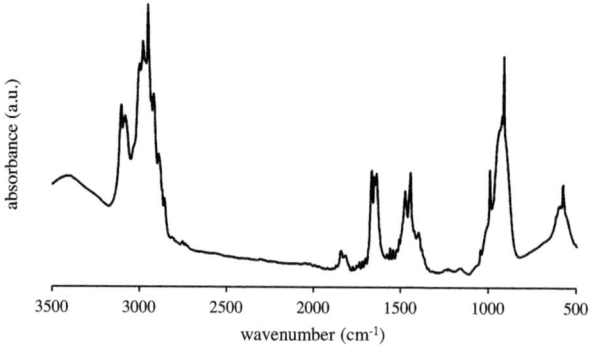

Figure 44. Transmission IR spectrum of the gas phase during the grafting of $Ti(O^iPr)_4$ to Ti_{mono}, revealing the formation of propene at 140°C.

From these experimental observations, it is our hypothesis that Ti(O^{i}Pr)$_4$ mainly reacts with the Ti-Cl surface sites of Ti$_{mono}$, resulting in the formation of dimeric Ti-O-Ti surface sites as shown in Scheme 8. No definitive statement can be made on the precise molecular structure, *e.g.* whether the two Ti-atoms are linked *via* an oxo-bridge and a μ^2-bridging propoxy group (dimer a), or only one oxo-bridge (dimer b). However, DFT-calculations indicate that upon computational optimization dimer b with only one oxo-bridge is more stable (*viz.*, ≡TiO-Ti(O^{i}Pr)$_3$, see scheme 8).

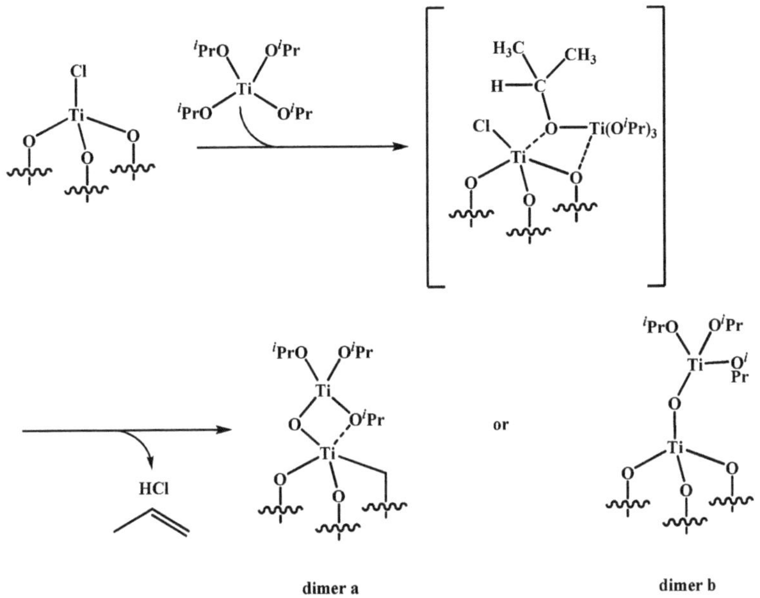

Scheme 8. Reaction of the Ti-Cl species on the Ti$_{mono}$ material with Ti(O^{i}Pr)$_4$.

Nevertheless, as mentioned before, the synthesis of the Ti_{mono} also creates ≡Si-Cl surface sites which are probably responsible for the fact that the Ti-loading increases by a factor higher than 2, through the formation of additional tetrahedral Ti^{IV}-sites (see Scheme 9).

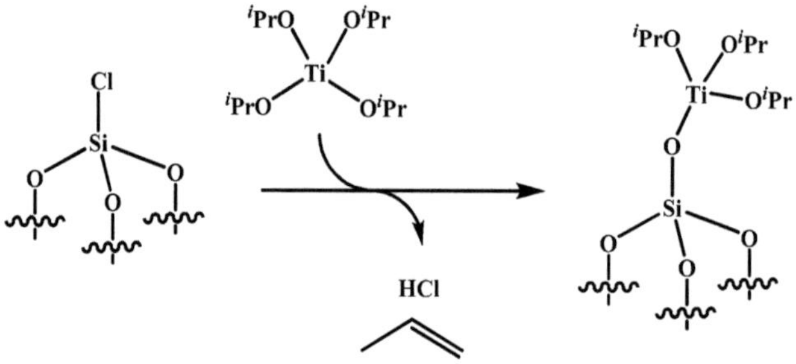

Scheme 9. Reaction of Si-Cl bonds on Ti_{mono} with $Ti(O^iPr)_4$.

Further characterization of the Ti-O-Ti species was completed using X-ray spectroscopy. XANES analysis indicates a small pre-edge feature (Figure 45), usually attributed to tetrahedral coordinated Ti atoms (Figure 45) due to the electronic 1s → 3pd transition. The EXAFS analysis shows no evidence for a scatterer at 2 Å (typical distance for a Ti-Cl bond as observed for Ti_{mono}), in line with bulk analysis where no Cl could be found by ICP OES. Even though it was possible to fit a model with a Ti-Ti distance of 3.07 Å (Figure 46) EXAFS only provides an averaged picture of all Ti surface sites. This is also reflected by the coordination numbers in table 2 which are a mixture of monomeric and dimeric Ti-sites with *iso*-propoxide and ≡Si-O-ligands of the silica matrix. Due to the presence of a small amount of isolated sites (see above), in addition to the dimer sites, a precise molecular structure determination by EXAFS is impossible. The fitted Ti-Ti distance of 3.07±0.02 Å appears to be in between the predicted value of

2.8 Å for dimer a and 3.46 Å for dimer b (see scheme 8). The true nature of the dimeric Ti species might also be dependent on the local environment provided by the silica matrix.

From the characterization it is concluded that $Ti(O^iPr)_4$ reacts with the Ti-Cl bonds on the Ti_{mono} material, yielding a material which comprises dimeric TiOTi species.

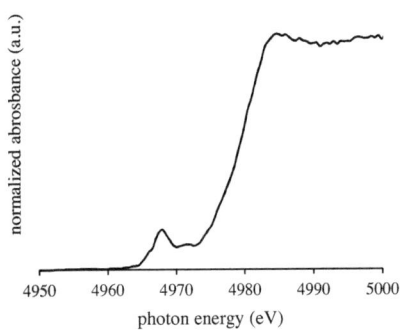

Figure 45. Ti K-edge XANES of the Ti_{dimer} material.

distances	N	Distance R (Å)	Debye-Wallis-factor σ (Å²)	R-value	independent points
Ti-O	4.5	1.87 ± 0.02	0.014 ± 0.0012	0.026	11
Ti-Si	1.5	2.78 ± 0.03	0.01		
Ti-C	3	3.34 ± 0.02	0.005		
Ti-Ti	1	3.07 ± 0.02	0.001		

Table 2. EXAFS fitting results with their uncertainties. Variables which were fixed during fitting are grey colored.

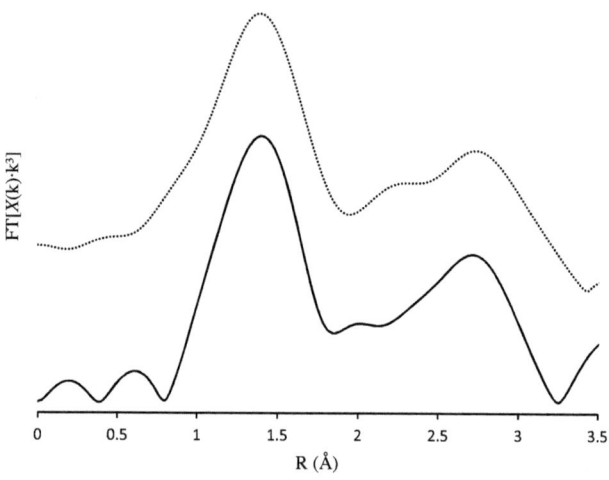

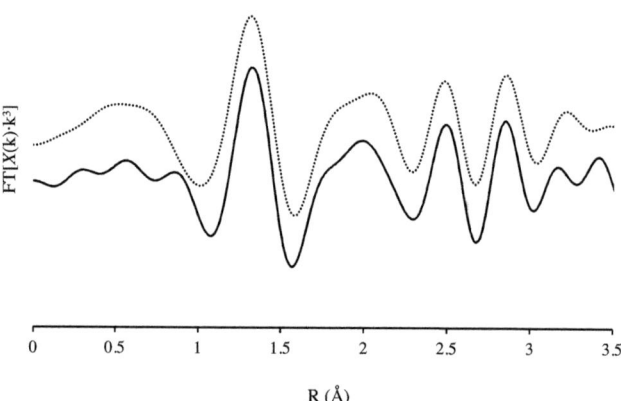

Figure 46. Ti EXAFS spectra (solid line) and fits (dotted line) (top panel: FT magnitude; lower panel: real part; fitting parameters are summarized in Table 1).

4.3 Conclusions

In this contribution, the synthesis and characterization of Ti-O-Ti dimer sites on silica are reported. The metal surface species are well dispersed on silica through the reaction of $Ti(O^iPr)_4$ with site-isolated $\equiv$TiCl species which are prepared by grafting $TiCl_4$ onto silica with a thermal post-treatment at 450°C. Even though no precise molecular structure could be concluded, the disappearance of the Ti-Cl bonds as observed by ICP, EXAFS and UV-Vis indicates the formation of Ti-O-Ti moieties on the silica surface upon grafting of $Ti(O^iPr)_4$. Thus, the described procedure is a convenient synthesis approach for a wide range of bimetallic surface sites which are of broad interest to the field of heterogeneous catalysis. The prepared materials can be used to investigate synergic effects of two metals in the catalytic performance in different chemical reactions in comparison to their isolated monomeric analogues.

Part III

Kinetic Investigations

5 Kinetic Experiments with *tert*-Butyl Hydroperoxide

In this chapter, the kinetic activity of monomeric Ti^{IV}-sites (T450 or Ti_{mono}) was compared to the activity of dimeric Ti^{IV}-sites (Ti_{dimer}) in olefin epoxidation with TBHP under continuous-flow conditions. Only a minor difference in turnover frequency was observed, being slightly higher for the monomeric Ti^{IV}-sites. Models with different ring-strains stemming from the silica support were computationally investigated. The model with the least ring-strains showed the greatest agreement with the experimentally obtained energy barriers best.

5.1 Introduction

Epoxides are one of the most important chemical intermediates in industry[36, 64]. Presently, all large catalytic epoxidation processes are mediated by Ti^{IV}-based heterogeneous catalysts, *e.g.* the Styrene Monomer Propylene Oxide (SMPO) technology[33] or Hydrogen Peroxide Propylene Oxide (HPPO) technology)[2]. For instance, the catalyst used in Shell's SMPO process is $TiCl_4$ grafted onto silica[33] whereas TS-1[2] is used in the HPPO process. The active center of both catalysts is connected via oxygen bonds to the silica matrix and it is general assumed that high activity is associated with site-isolated tetrahedral Ti^{IV} species[233, 272, 288b, 326, 327, 328]. It was reported in literature that avoiding the formation of octahedrally coordinated TiO_2 species is of importance than the precise titanium dispersion[335]. As long as tetrahedral Ti^{IV}-sites are protected against nucleation, high epoxidation activity could be observed. Additionally, Scott *et al.* synthesized oxo-bridged Ti-O-Ti dimers on silica surfaces which are very efficient for catalytic epoxidations with *tert*-butylhydroperoxide[271].

The true active species in the catalytic cycle however, is still a point of discussion[33, 254-256]. In order to elucidate this species in more detail silica grafted $TiCl_4$ thermally post-treated at 450°C can be considered as a model catalyst exhibiting a similar active species as in the aforementioned industrially applied catalysts. By computational modelling of the rate determining step, the transfer of the oxygen onto the olefin, conclusions can be made about the active site in the solvent free epoxidation of different olefins with TBHP. One example is the increased stability during this process when the material is heated to 450°C instead of 250°C as mentioned in chapter 2.

In this chapter, different kinetic experiments are conducted to draw further structure-activity relationships demonstrating the suitability of T450 as a model catalyst. Moreover, the activity of its site-isolated $\equiv$TiOR sites are compared with oxo-bridged dimer sites of the T_{dimer} catalyst from the previous chapter, in order to access the influence of the molecular environment of the Ti^{IV} on its catalytic performance.

5.2 Catalytic Epoxidation with *tert*-Butyl Hydroperoxide

In order to compare the catalytic activity of the Ti_{mono} and Ti_{dimer} materials, the solvent-free olefin epoxidation with *tert*-butyl hydroperoxide as oxidant was taken as a model reaction under continuous flow conditions (see Experimental section).

As a proof of principle, a set of kinetic experiments with cyclohexene was performed, varying the contact time and temperature (Figure 47). An Arrhenius activation energy of 13 ± 1 kcal · mol^{-1} could be measured in the temperature range 60-80 °C (see inset of Figure 47).

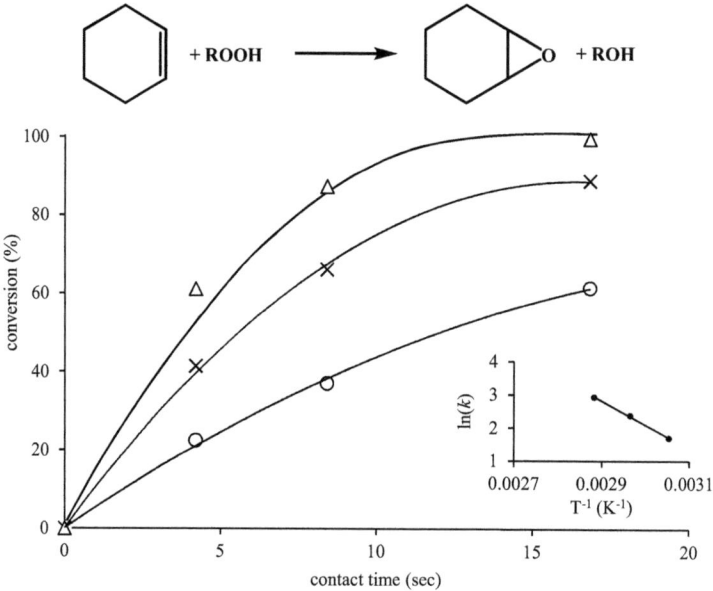

Figure 47. Epoxidation of cyclohexene with Ti_{mono} (8 µmole Ti); inset shows the Arrhenius plot.

The reaction of the ≡TiCl surface species with the *tert*-butyl-OOH oxidant resulted in the formation of ≡Ti(η^2-ROO) species upon removal of chlorine (see scheme 10 and figure 48). Those titano-peroxo species are active in olefin epoxidation presumably by transfering the distal peroxo O-atom to the substrate while the proximal O-atom preserves the ligation to the metal center.

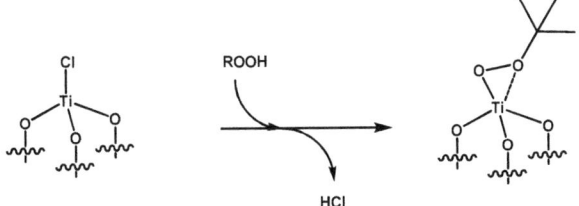

Scheme 10. Reaction of the ≡Ti-Cl species with ROOH to the epoxidizing ≡Ti(η^2-ROO) species.

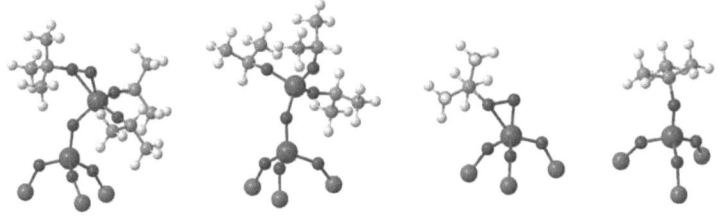

Figure 48. Models for (from right-to-left) the ≡TiOR and ≡Ti(η^2-ROO) species on the Ti$_{mono}$ material, and ≡TiO-Ti(O^{i}Pr)$_3$ and ≡TiO-Ti(O^{i}Pr)$_2$(η^2-ROO) species on the Ti$_{dimer}$ material.

We computed the barrier of the rate-determining oxygen-transfer step for various small cluster models of the active Ti-site, revealing that the introduction of ring-strain decreases the activation barrier (see Figure 49). Comparing the computed barriers for the three models with the experimental

Arrhenius energy of 13 kcal mol^{-1} suggests that the active site on the Ti$_{mono}$ material is most likely a fairly unstrained tripodal Ti^{IV}-species.

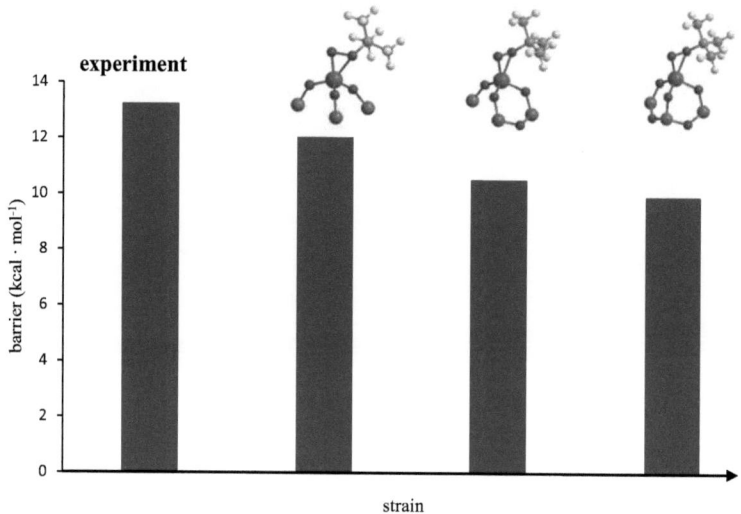

Figure 49. Comparison of the experimental activation energy for cyclohexene (*i.e.* 13.2 ±1 kcal mol^{-1}) with the computed activation barriers, based on tripodal Ti^{IV}-sites without strain (12.3 kcal mol^{-1}), slight strain (10.5 kcal mol^{-1}) and significant strain (9.9 kcal mol^{-1}).

The experimental barriers of a range of other (cyclic) olefins were also reproduced by the small model for the unstrained tripodal active Ti^{IV}-site, under solvent-free conditions and in acetonitrile (computationally modeled with polarizable continuum model; experimental [olefin] = 100 mM in case of acetonitrile as solvent) as shown in Figure 50.

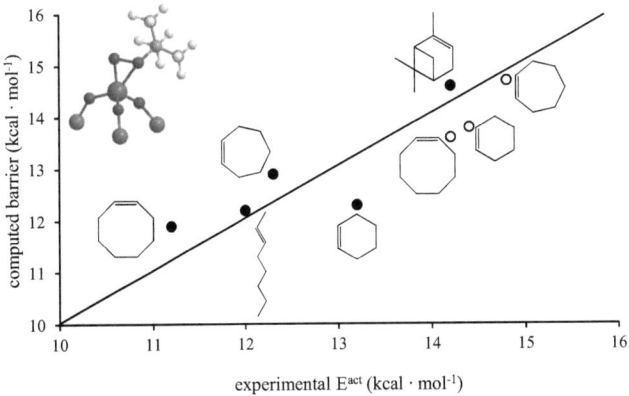

Figure 50. Comparison of the experimental Arrhenius activation energy obtained for various substrates with the computationally predicted barrier (● refers to solvent free systems, o to reactions performed in acetonitrile [olefin] = 100 mM).

Subsequently the Ti$_{dimer}$ material was tested in solvent-free olefin epoxidation as well. Surprisingly, no significant difference in activation energies could be observed in the investigated temperature range. In fact, the initial TOFs were approximately 40 % lower for the Ti$_{dimer}$ than the Ti$_{mono}$ material for all investigated substrates (*i.e.*, cyclohexene, -heptene, -octene, 2-octene and α-pinene). O-transfer from the distal peroxo O-atom in the ≡TiO-Ti(O^{i}Pr)$_2$(η^2-ROO) species (see Figure 48) was computationally investigated and found to have a similar barrier than transfer from the ≡Ti(η^2-ROO) isolated species. It is emphazied that the ROOH oxidant is preferably reacting with the none-surface-bound Ti center, displacing one of the alkoxy ligands (see Scheme 11). Additionally, protonation and dissociation of the Ti-O-Ti linkage was found to be less favorable than protonation and decoordination of an alkoxyl ligand, due to the lower negative charge and steric congestion. As such, only the non-surface bound Ti-site is kinetically active explaining the lower intial TOFs. No evidence for a synergetic effect of two Ti atoms in close proximity could be found for the dimeric sites as previously suggested[325]. These findings are in line with the work of Brutchey *et al.* reporting that isolated tetrahedral Ti^{IV}-sites are more important than a precise titanium dispersion[335].

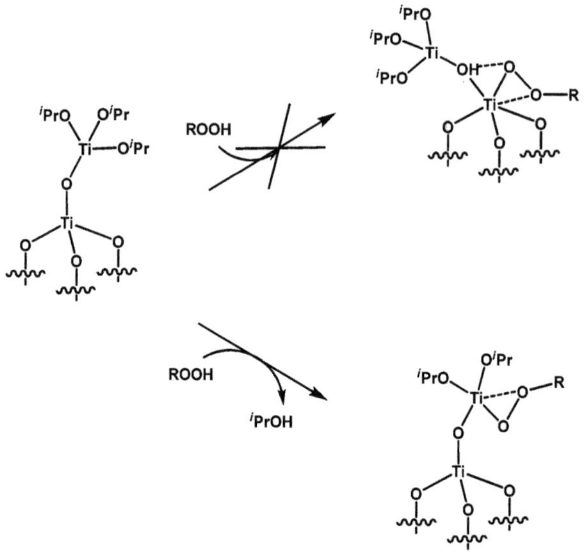

Scheme 11. Reaction of the ROOH oxidant with the ≡TiO-Ti(O^{i}Pr)$_3$ dimer sites.

Based on the experimental and computational observation, a catalytic cycle is proposed for solvent-free epoxidation of olefins (Figure 52). Starting from the active peroxo Ti species, the distal oxygen is transferred onto the olefin. Since no other reaction products except *tert*-butanol are detected by GC-FID the oxygen transfer is thought to be the only available reaction pathway in this temperature range. Moreover, the cleavage of the *tert*-butyl group from the proximal oxygen can be ruled out as no indication for oxidant decomposition could be observed. This is further supported by the fact that the UV-Vis spectra of T450 before and after operation showed no difference apart from a slight blue-shift of the maximum due to ligand replacement (figure 51). In order to close the catalytic cycle the butoxy ligand needs to be replaced by TBHP. Considering the higher pKa of TBHP (pKa ≈ 12) compared to *tert*-butanol (pKa = 17) the substitution of butoxyl group might not only be due to its weaker coordination to the Ti but also due to its higher proton affinity.

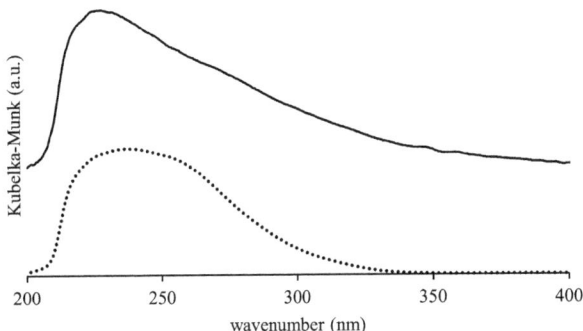

Figure 51. UV-Vis spectra of as-prepared T450 (dotted line) and after operation in solvent-free epoxidation and subsequent drying (solid line).

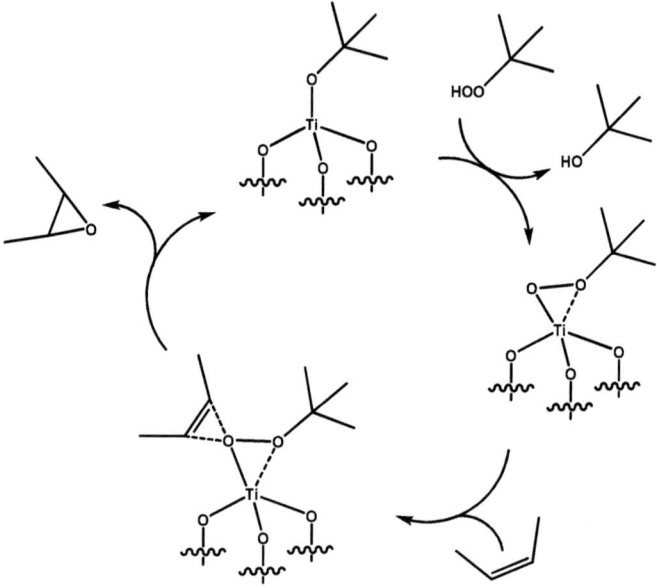

Figure 52. Possible Eley-Rideal mechanism for the heterogeneous epoxidation with *tert*-Butyl hydroperoxide under water-free conditions.

5.3 Conclusions

TiCl$_4$ grafted to amorphous silica with a thermal posttreatment at 450°C is active in the solvent-free epoxidation of different olefins. The detailed structural analysis in chapter 2 in combination with experimental and computational methods support the hypothesis that a fairly unstrained Ti^{IV}-site is the active species. Additionally, the experimentally and computationally obtained kinetic activation barriers for different olefin not only support this hypothesis but also confirms the oxygen-transfer to be the rate limiting step.

The activity of the obtained dimer sites was compared with ≡TiOR sites for the epoxidation of olefins with *tert*-butyl hydroperoxide, however no synergetic effect could however be observed. Nevertheless, the described synthetsis procedure is a convenient approach to a wide range of bimetallic-sites and is of broad interest to the field of heterogeneous catalysis[328]. Additionally no leaching of Ti was detected by ICP OES supporting the hypothesis based on computational methods that the Ti-O-Ti bond stays intact under epoxidation conditions.

6 Kinetic Experiments with Hydrogen Peroxide

In the final part, the stability of T450 is demonstrated during cyclooctene epoxidation with aqueous hydrogen peroxide. Different kinetic experiments are performed in order to gain insight into the catalytically active Ti-species. At 60°C, an activation barrier for epoxidation of cyclooctene with H_2O_2 of 11.7 ± 1 kcal $\cdot$ mol^{-1} is observed as for the epoxidation with TBHP described in the previous chapter, i.e.11.2 ± 1 kcal $\cdot$ mol^{-1}. At reaction temperatures over 60°C, the activation energy increases to 23 ± 1 kcal $\cdot$ mol^{-1}. The same observation for the activation barrier was found in the presence of water (c_{water} = 1.1 M).

Kinetic experiments revealed a lower activity for T450 than TS-1 in H_2O_2 decomposition, being 20 ± 3 kcal $\cdot$ mol^{-1} for TS-1. However, by boiling T450 in water for one hour the activation barrier for hydrogen peroxide decomposition was increased to 24 ± 3 kcal $\cdot$ mol^{-1}. In addition to the increased barrier, a new peak in IR was observed at 950 cm^{-1} for T450.

6.1 Introduction

In 2012, BASF and Dow started the new HPPO plant in Thailand for the production of propylene oxide (PO) with hydrogen peroxide over the well-known TS-1 catalyst[2]. Its yearly capacity of 390 000 t is almost double as high as the capacity of the second biggest PO producing plant in Antwerp[2]. In comparison to other technologies, this process is producing less waste, has a lower demand in energy[217] and is not couple to the production of another product as in Shell's SMPO process[263]. The heart of this process is the heterogeneous catalyst titaniumsilicate-1 having a pore diameter of 5.5 Å and tetrahedrally coordinated Ti in the MFI structure[257]. DTG analysis[217], UV-Vis[264,247], IR[244], Raman[245,340] and X-Ray absorption[242,341] spectroscopy support the hypothesis that the Ti is incooperated into the defect sites of the silica network. By neutron diffraction, Lamberti *et al.* suggested four positions in the MFI-structure (T_6, T_7, T_{10}, T_{11}, see figure 53) preferably occupied by titanium[218].

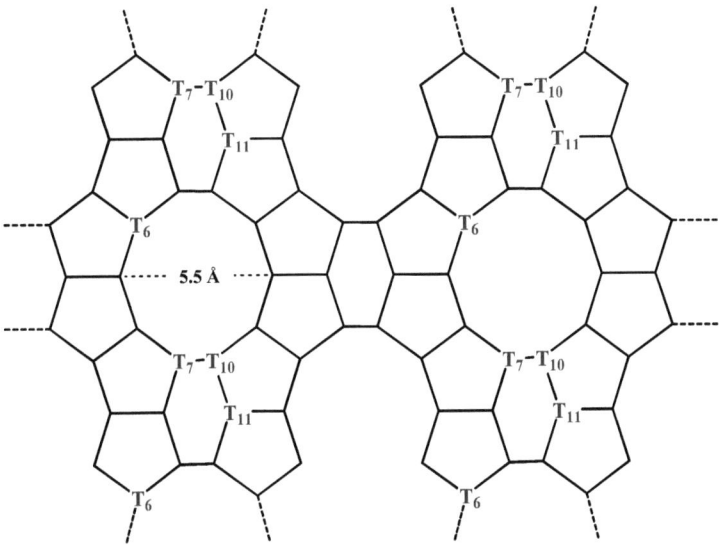

Figure 53. Cross-section perpendicular to the 5.5 Å pores in the MFI structure. Site which are preferably substituted by titanium atoms are in red.

Even though the Ti site in TS-1 is well characterized the active peroxo-titanium species is still under debate. TS-1 turns yellow when it is exposed to aqueous hydrogen peroxide but losesits color again upon dehydration overnight[217]. The same authors demonstrated the reversibility of this process by EXAFS and proposed the existence of two different peroxo-species. Prestipino and co-workers supported those finding by exposing dehydrated TS-1 to a moisture-free hydrogen peroxide source[342]. No color change was observed unless the TS-1 exposed to the peroxide source was also brought into contact with water indicating a crucial role of the H_2O molecule in the Ti-peroxide structure.

In order to elucidate the active peroxo-Ti species in more detail silica grafted $TiCl_4$ thermally post-treated at 450°C can be considered as a model catalyst exhibiting a similar active species as in the before mentioned industrially applied catalysts. By characterizing the molecular structure of the active species in T450 and comparing the kinetic performance of it to the one of TS-1, conclusions can be drawn about the kinetics and molecular strucutres involved in the reaction mechanism.

Recently, Hammond *et al.* reported a highly active and selective W/Zn/SnO2 catalyst for the epoxidation of cyclooctene[174]. Taking into account the price difference of the two metals precursor (WO_3 ca. 220 CHF per kg and $TiCl_4$ ca. 100 CHF per kg at Sigma-Aldrich) Ti catalyst is still an economical alternative to tungsten-based systems. Moreover, it has been reported that toxic long-term effects of tungsten exposure can cause severe health issues[342].

In this chapter, the well-characterized material T450 is kinetically investigated in the olefin epoxidation with hydrogen peroxide and the decomposition of the latter. The influence of water and reactivity between 40 and 80°C is studied as well.

6.2 Catalytic Epoxidation with Hydrogen Peroxide

In order to test the activity and stability of silica grafted $TiCl_4$ thermaly posttreated at 450°C (T450) in epoxidation with aqueous hydrogen peroxide, a continuous flow experiment were conducted. During two weeks on stream no significant decrease could be detected. Additionally, one batch and one semi-batch experiment according to the procedure reported by Hammond for their $W/Zn/SnO_2$ system were compared[174]. In the two different experiments, one of them was conducted as reported and in the other, hydrogen peroxide was slowly added during a period of 4 hours due to literature reports stating that successive addition (semi-batch) of hydrogen peroxide decreases its decomposition[343]. In Figure 54, T450 yields over 40% cyclooctene within two days when the oxidant hydrogen peroxide is added successively. However, a lower yield is obtained when all the hydrogen peroxide is present from the start. Due to the stability of T450 in the continuous flow experiment (Figure 55) the influence of gradual hydrogen peroxide addition can only be explained by the decomposition of the oxidant which will be discussed in detail later. Additionally, no leaching could be detected by ICP even though a small increase of isolated silanol groups was observed by IR spectroscopy (Figure 56) which is probably due to opening of strained siloxane bridges.

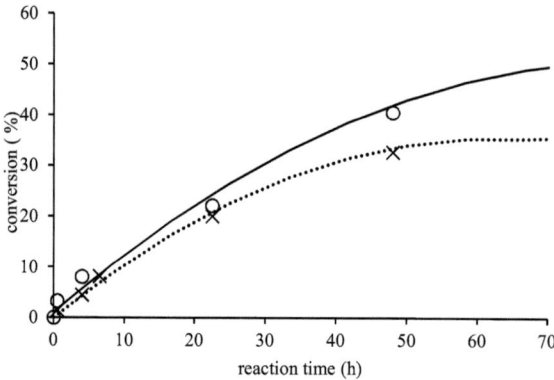

Figure 54. Batch experiments, reaction conditions: Cyclooctene in 1-butanol (1.0 M), olefin/H_2O_2 molar ratio = 1.0, 80 °C, 0.15 mol % Ti (circles: successive addition of H_2O_2).

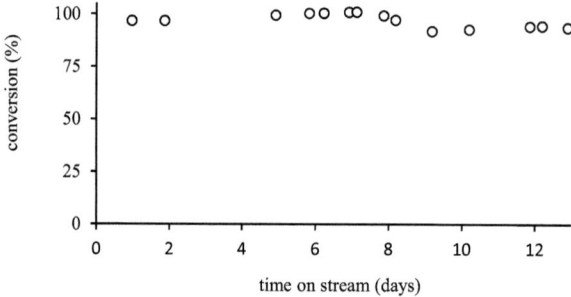

Figure 55. Epoxidation of cyclooctene under continuous flow at 0.05 ml/min and 60°C in dioxane/cyclooctene 2:1 with 100 mM H_2O_2;

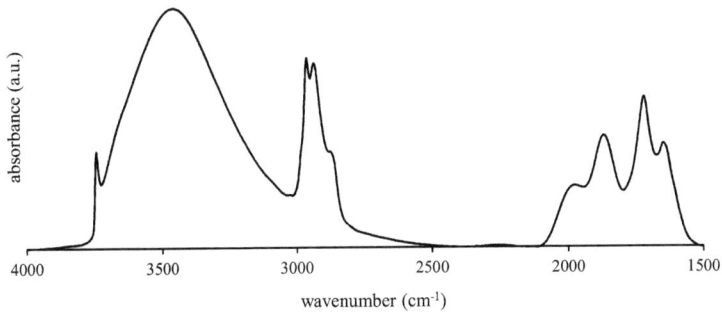

Figure 56. IR spectrum of T450 after 4 weeks on stream for epoxidation in dioxane/cyclooctene (2:1) with 100 mM hydrogen peroxide.

In order to study the behavior of T450 in cyclooctene epoxidation with hydrogen peroxide, continuous flow experiments were conducted at temperatures ranging from 40-80°C (Figure 57). Below 60°C an activation energy of 11.7±1 kcal · mol^{-1}, which is very similar to the activation energy for the epoxidation with TBHP (11.2±1 kcal · mol^{-1}), was found. Above 60°C, the activation energy rises to 23±1 kcal · mol^{-1} suggesting a different mechanism with a higher temperature dependence. In addition, the formation of molecular oxygen in the catalyst bed was observed for temperatures between 70 and 80°C.

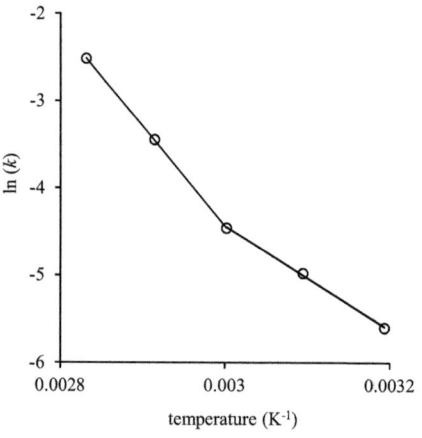

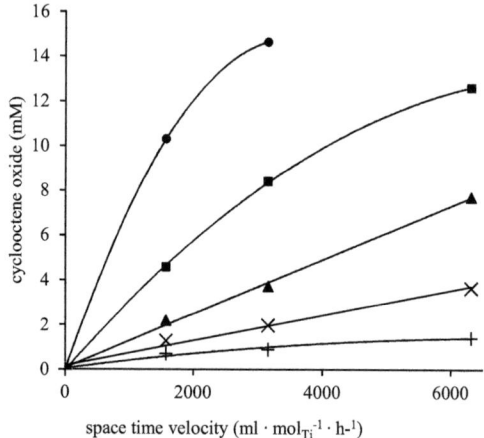

Figure 57. Bottom: Epoxidation in dioxane/cyclooctene (2:1) with 100 mM H_2O_2 (10 min contact time) at 40(+), 50(×), 60(▲), 70(■) and 80(●) °C; Top: Corresponding Arrhenius-plot with first-order-rate constants k.

In order to avoid interference with hydrogen peroxide decomposition at higher temperature, epoxidation of allylic alcohol were conducted at carefully adjusted temperatures of 55, 60 and 65°C. The reactions were performed in a mixture of dioxane/cyclooctene (2:1) with increasing water content (Figure 58). The activation energies and Arrhenius-factors (table 3) are of first-order as reported in literature[344] and increase with higher water content (figure 59). Since the change in activation energies is correlated to the water concentration it seems like two (or more) different mechanisms are responsible for allylic alcohol epoxidation. When *tert*-Butyl hydroperoxide is used as an oxidant with a water concentration is high (0.5 M), no activity could be detected by GC/FID (table 3, last entry) which indicates that the presence of water impedes the other mechanism which is probably due to blocking of the active site as previously reported for microporous materials[345]. On the other hand, the Arrhenius-factor increases (figure 59) at higher water concentrations. A higher Arrhenius-factor implies an increased number of successful reactions per active site. This points towards the fact, that the active species in the reaction mechanism with an activation energy of 23 ± 3 kcal $\cdot$ mol^{-1} is formed upon addition of water.

oxidant	water concentration (M)	activation energy (kcal · mol^{-1})	arrhenius factor (s^{-1})
^{t}BuOOH	0	7±3	4·10^4
H$_2$O$_2$	0.2	14±3	1·10^8
H$_2$O$_2$	1.1	23±3	7·10^{13}
^{t}BuOOH	0.5	inactive	inactive

Table 3. Epoxidation of allyl alcohol with T450 between 55 and 85 °C from Figure 58 and 59.

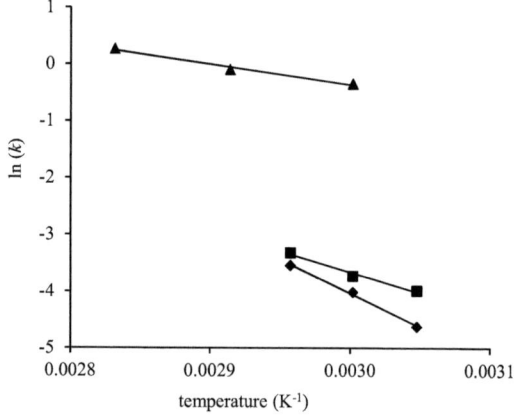

Figure 58. Arrhenius-plots of epoxidation of allyl alcohol at different water concentrations (▲ = 0 M; ■ = 0.2 M; ♦ = 1.1 M);

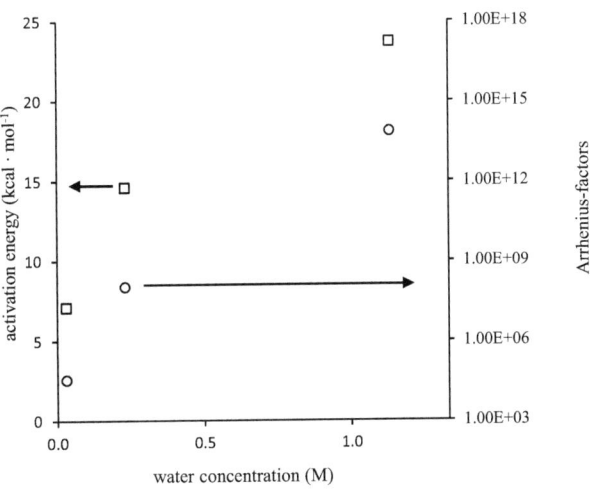

Figure 59. Corresponding Activation energies (squares) and Arrhenius-factors (dots)

The mechanisms of epoxidation with hydrogen peroxide is controversly debated in literature where a whole range of mechanisms has already been proposed for different epoxidation systems[220,228,229,230,,335,336]. Amongst those, the mechanism of Don Tilley involves the contribution of a proton acceptor (Me$_3$E-O-group, E= Si, Ge, Sn). Following his mechanism we suggest a similar Langmuir-Hinshelwood-mechanism for our system (figure 60). A crucial point might be the second deprotonation of a H$_2$O$_2$ molecule coordinated to Ti as shown in the transition step in figure 60. We suggest that this species could be responsible for the increase observed in activation energy at higher temperatures or water concentrations. As pointed out by Lamberti et al.[217] the presence of water decreases the pH of this system and even though the activation barrier for the formation of the doubly deprotonated peroxo species might be higher, it is very probable that it is more active in epoxidation than the hydroperoxo species. This point will be discussed in more detail in the next section.

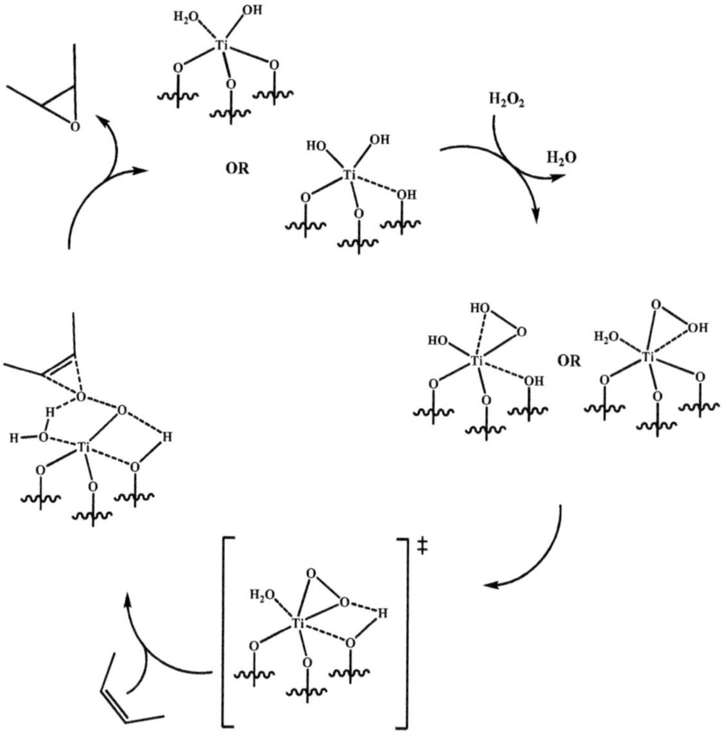

Figure 60. Possible Langmuir-Hinshelwood mechanism for the heterogeneous epoxidation with aqueous H_2O_2.

6.3 Decomposition of Hydrogen Peroxide

In order to obtain a better understanding of the chemistry of hydrogen peroxide in the presence of heterogeneous Ti-catalysts, decomposition experiments with T450 and TS-1 were conducted in the temperature range between 60 and 80°C. Samples were quantified by ceriometry with a 1M cerium sulfate solution in diluted sulfuric acid. Pure Aerosil200 from Evonik showed only negligible activity in H_2O_2 decomposition as (Figure 61, black points). However, after silylating Aerosil200 no activity in decomposition could be detected. T450 was thoroughly washed with butanol before it was used for decomposition experiments in order to avoid the interference of Cl or HCl, formed upon the reaction of surface Ti-Cl bonds with the HO-group of butanol. T450 decomposed H_2O_2 slightly faster than Aerosil200. Since the activity of T450 did not significantly change upon silylation, this observation indicates that H_2O_2 is decomposed faster by the Ti-sites in comparison to the silanol groups. For TS-1 a higher activation energy was measured but due to an increased Arrhenius-factor by the power of 10 the material was more active in H_2O_2 decomposition. No decrease in Ti-loading of the recovered catalysts could be detected by ICP OES.

catalyst	Not boiled	boiled	Not boiled	boiled
	Arrhenius $[l \cdot (g \cdot s)^{-1}]^a$	*Arrhenius* $[l \cdot (g \cdot s)^{-1}]^a$	*Activation energy* $[kcal \cdot mol^{-1}]$	*Activation energy* $[kcal \cdot mol^{-1}]$
TS1	10^{10}	10^9	20±3	19±3
T450[a]	0.7	10^9	10±3	24±3
silyated T450[b]	0.004	---	10±3	---
Aerosil200©	---	---	10±3	---
silylated Aerosil200©	inactive	---	inactive	---

Table 4. Kinetic results for the decomposition of hydrogen peroxide between 55 and 85 °C determined by ceriometry; [a]normalized to the amount of Ti; [a] washed with tert-butanol; [b] washed with water before silylation.

In further kinetic studies, the catalysts were first boiled for 1h in water before their decomposition activity was monitored (table 4 and figure 61). The boiling pretreatment showed no significant effect on TS-1 but it had a tremendous one on T450. The Arrhenius-factor and activation energy for T450 increased up to a similar level as for TS-1. This effect was not observed upon boiling Aerosil200©. With IR a small peak around 960 cm^{-1} was detected after boiling T450 in water which is also present for TS-1 (Figure 61 b)). We believe that this vibration is due to a Si-OH group which is coordinated to a Ti atom. Uncoordinated Si-OH groups vibrate above 1000 cm^{-1} and a Si-OH vibration with a coordinated Ti atom makes the oxygen "heavier" and thereby lowers the frequency of this vibration. A coordinated Si-OH group can be formed due to hydrolysis of Si-O-Ti group during boiling and is significantly more acidic than an isolated silanol group making it more active in H_2O_2 decomposition. The O-H bond could not be detected by IR or NMR spectroscopy probably due to very broad signals.

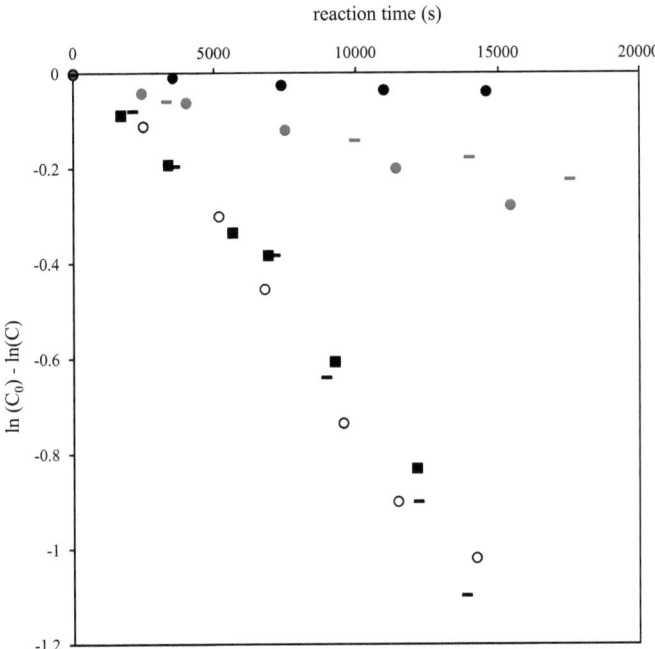

Figure 61a). Decomposition of H_2O_2 at 80°C of different materials: ● = Aerosil200©, ● = T450 washed with ^{t}BuOH, — = silylated T450; ■ = T450 boiled in water; o = as-prepared TS-1; - = TS-1 boiled in water.

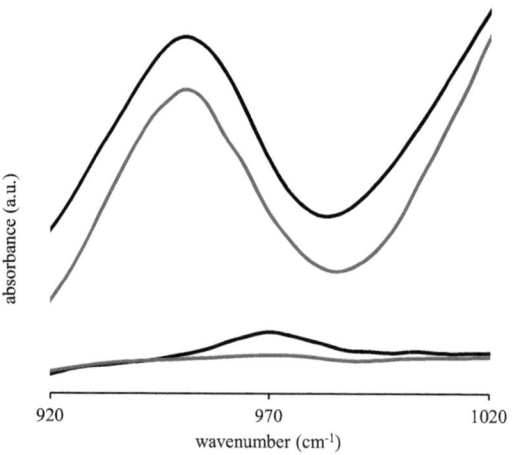

Figure 61 b). IR spectra of catalysts before (grey) and after (black) boiling in water; top: TS-1; bottom: T450

The OH-region in IR indicates an increased concentration of networked O-H groups after boiling or using the catalysts for hydrogen peroxide decomposition due to absorbed water or H_2O_2 (figure 62). For the T450 materials the amount of isolated silanol groups after washing in pure butanol increased slightly because a small amount of siloxane bridges was opened and ICP disclosed a small but significant loss of 0.2 wt% surface Ti. The increase in networked silanol groups after boiling is probably due to the hydrolytic opening of Ti-O-Si groups resulting in Si-OH groups coordinated to a Ti species. Lamberti *et al.* observed a similar effect when they decreased the number of defects sites by a higher Ti incorporation and therewith the number of networked silanols[217]. In line with their work, boiling of T450 in water opens Ti-O-Si bonds increasing the amount of hydrogen bonding network, resulting in peak around 960 cm^{-1} and an increased activity in hydrogen peroxide decomposition. To best of our knowledge, this is the first time that the activity in H_2O_2 decomposition could be directly correlated with an IR vibration around 960 cm^{-1}. Most reports correlated this peak with the activity in various reactions, such as in epoxidation of 2-cyclohexenol[336] or oxidation of benzene to phenol[335]. This is rationalized by the fact that the peak is proportional to the Ti-loading[326]. Moreover, taking into account that three of the four preferred Ti-positions in TS-1 (T7, T10 and T11) are adjacent to each other implies that a significant number of Ti atoms could be close to a Si vacancy or silanol nest[217]. The close proximity of those two species increases the number of silanols coordinated to Ti atom associated with the IR peak around 960 cm^{-1}. The increased acidity of the coordinated Si-OH group could explain the higher activity in H_2O_2 decomposition.

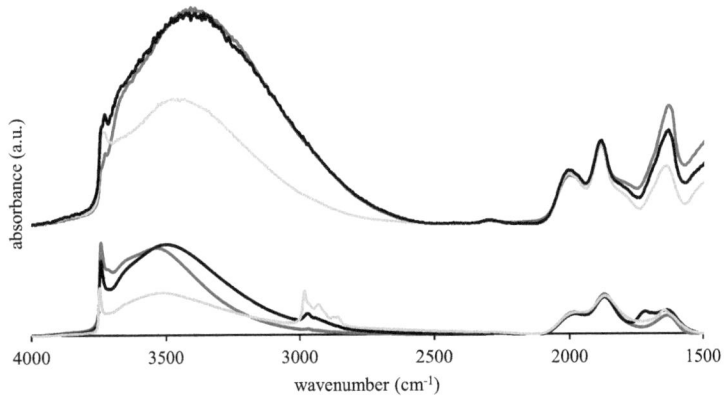

Figure 62. IR spectra of different catalysts*: T450 washed with ^{t}BuOH (light grey line, bottom); T450 boiled in water (dark grey line, bottom); T450 washed with ^{t}BuOH and used in decomposition experiments (black line, bottom); as-prepared TS-1 (dark grey line, top); TS-1

boiled in water (light grey line); TS-1 used for decomposition experiments (black line, top); *TS-1 catalysts with an offset

UV-Vis spectroscopy showed no change after boiling or washing the catalysts with water (Figure 63 and 64). Only after exposing to aqeous H_2O_2 the as-prepared and boiled TS- a shoulder was observed between 250 and 300 nm associated with a η^2-peroxo Ti species responsible for the yellow color after exposure[217,342]. No decrease in Ti loading upon boiling in water was detected by ICP, neither for TS-1 nor for T450.

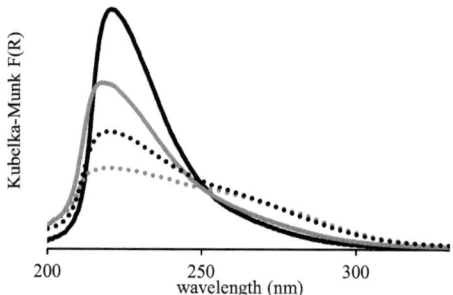

Figure 63. UV-Vis spectra of used catalysts used for decomposition of H_2O_2 Left: as-prepared TS-1 (solid black line), TS-1 after boiling (solid grey line), after decomposition experiment (dotted black line), after boiling and after decomposition experiment (dotted grey line).

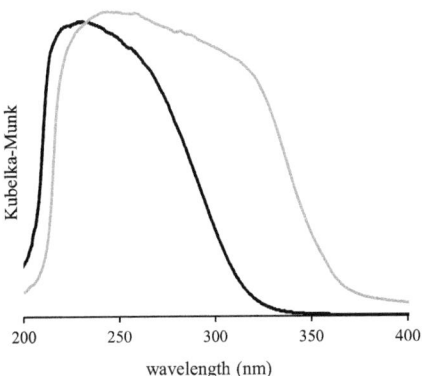

Figure 64. T450 boiled in water (black) and T450 washed with n-Butanol (grey)

Similar to UV-Vis and ICP, XANES (Figure 65) and EXAFS analysis (Figure 66) indicated no change in the active titanium center of the T450 before and after using the catalyst in H_2O_2 decomposition. The EXAFS fits were done with a Ti center coordinated with five oxygens because a first-shell analysis gave 5-6 atoms at a distance of 1.9 Å and 3 silicons at a distance of 2.8 Å in the second shell. The two fits were done in R-space from 1.2-2.8 Å and the fitting parameters can be found in table 5 and 6.

The fit for the butanol washed T450 (Figure 66, top spectra) was done with two additional carbon scatterer at a distance of 2.8 Å which resemble the carbons in a-position to the oxygen of the butanol ligands. Only E_0 and the R-values were allowed to be varied. The Debye-Wallis-factors were fixed to 0.0015 for carbon, 0.0023 for silicon and 0.0045 for oxygen according to their number in the model fitted. For the oxygen, the σ^2-values are slightly higher since different Ti-O distances (O-*Si* and O-*C*) are averaged out to one oxygen distance in this model due to the limited number of independent points. Moreover, some butanol molecules are probably coordinated to the Ti resembling a third Ti-O distance which is probably longer than 2 Å and resulting in Ti coordination sphere with more than 4 oxygen atoms as observed be XANES and the first-shell-analysis.

The fit for the T450 which was used for kinetic experiments in hydrogen peroxide decomposition was fitted without the carbon scatterer as above. The Debye-Wallis-factor of the oxygen is slightly too low which is due to a higher concentration of octahedrally coordinated titanium sites after decomposing hydrogen peroxide in water. However, it was still possible to fit 3 silicon atoms to the second sphere which showed that the Ti does not leave the silica surface during exposure to a 2.5 M H_2O_2 solution at 80°C for 6h. This confirms the stability achieved by our post-synthetic vacuum heat treatment.

The fact that both fits involve three silicon atoms in the second coordination sphere indicates that the Ti stays bound to the silica surface throughout the decomposition experiments.

T450 used in H₂O₂ decomposition   T450 washed with *t*-Bu

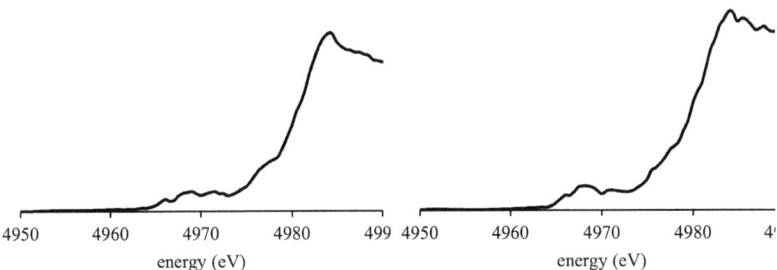

Figure 65. XANES of T450 washed with butanol (right) and used for hydrogen peroxide decomposition (left).

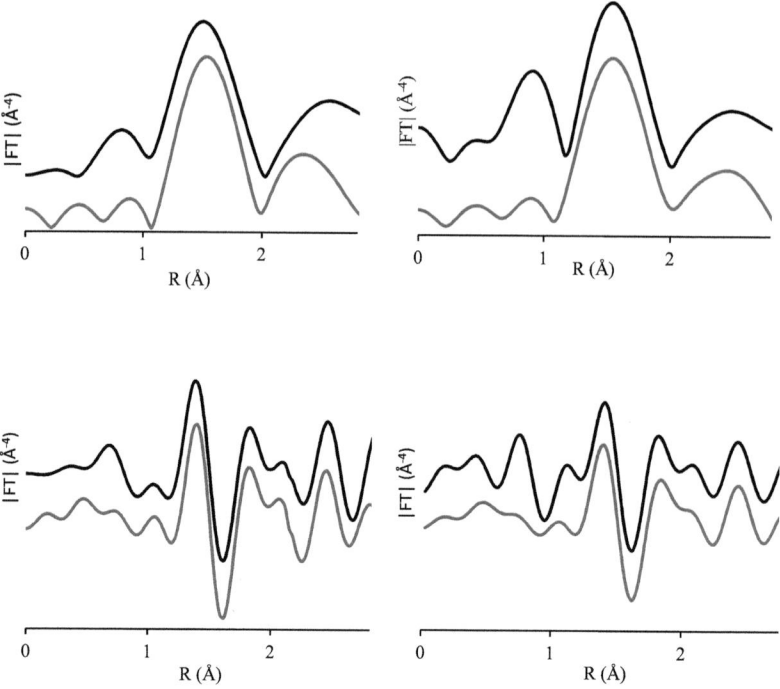

Figure 66. EXAFSspectra (black) and fits (grey) (right side: magnitude; left side; real part of Fourier-transformation) for *tert*-butanol washed T450 (bottom spectra) and T450 used for decomposition of H_2O_2 (upper spectra)

Scatter.	N	R	σ^2	Independ. points	Fitting range	R-factor	E_0
O	5	1.98±0.02	0.005		1.2-2.8 Å	0.03	5.9±3.8
Si	3	2.79±0.04	0.002	7			
C	2	2.76±0.14	0.002				

Table 5. EXAFS fitting data for T450 washed in butanol

Scatter.	N	R	σ^2	Independ. points	Fitting range	R-factor	E_0
O	5	1.96±0.02	0.001		1.2-2.8 Å	0.03	4.1±3.7
Si	3	2.79±0.03	0.002	6			

Table 6. EXAFS fitting data for T450 after H_2O_2 decomposition

6.4 Conclusions

It was shown by different kinetic experiments that T450 exhibits activity for olefin epoxidation with aqeous hydrogen peroxide. Below 60°C, it has a low activation energy of 11.7±1 kcal · mol^{-1} for cyclooctene while at temperatures above 60°C an activation energy of 23±1 kcal · mol^{-1} was obtained. Epoxidation with hydrogen peroxide in dioxane showed an increased activation barrier up to 23±3 kcal · mol^{-1} when water was added to the reaction solution (1.1 M). We attribute the increase in activation energy to the presence of a new active species (η^2-peroxo titanium) which is formed upon addition of water as reported in literature for TS-1[217,341]. Even though the activation barrier for the doubly deprotonated form of hydrogen peroxide is higher, this species is more active than the η^2-hydroperoxo species due to a higher Arrhenius-prefactor. The formation of this species was not only observed by the addition of water but also by increasing the reaction temperature which is rationalized by the entropically favored dissociation of the second O-H bond at higher temperatures. Moreover, similar activation barriers in epoxidation and decomposition with aqueous H_2O_2 by boiled T450 indicate a similar rate-determining step for both reactions. If the proton-acceptor ability of water is responsible for the formation of the doubly deprotonated peroxo-Ti species H_2O_2 decomposition could be avoided by adding catalytic amounts of mild Lewis acids which are unable to open epoxides. Consequently, the superior H_2O_2 selectivity of TS-1 would be due to the avoidance of water as a proton acceptor at the active Ti-sites[256, 346] and not due to blocking of the Ti-sites[345]. However, a deeper study of this phenomenon is necessary in order to confirm this hypothesis.

After boiling T450 in water its activity in H_2O_2 decomposition was similar to that of TS-1. A peak around 960 cm^{-1} was detected for TS-1 and water boiled T450 which was associated with a silanol coordinated to Ti and thereby increasing the amount of hydrogen bonding networks (figure 7) and the activity in decomposition (figure 57). The results support the hypothesis of Lamberti *et al.* that Ti atoms in TS-1 are incorporated in close proximity to a silanol nest (or Si defect site) resulting in silanol groups coordinated to Ti atoms[217].

No indication of leaching or change in the Ti-sites was observed by ICP, EXAFS, XANES or UV-Vis.

Part IV

Outlook

7 Outlook

In this work various materials prepared by chemical vapor deposition or grafting in liquid phase were characterized. It was shown that by increasing the post-treatment temperature to 450°C under moisture-free conditions the catalyst exhibits superior activity and stability. In Shell's SMPO process the catalyst is flushed with steam before the thermal post-treatment. This could hydrolyze the Ti-O-Si bonds of the formed monopodal surface species during grafting. Due to cleavage of Ti-O-Si bond, the surface Ti species may be subsequently able to move along the surface leading to the formation of TiO_x-particles[33]. A higher stability could be obtained by omitting the steaming step and directly treating the freshly-grafted catalyst at high temperature as it has been shown that by Cl-transfer to the silica surface multipodal surface species are formed. In future investigation, it would be of interest to know at what temperature a second Cl-atom is transferred to silica surface, leading from a bi- to a tripodal surface species.

It might be advantageous to prepare the same catalyst on a support with a higher surface density of silanol groups to achieve a higher metal loading. With an increased Ti surface concentration, not only the signal intensity could be increased but additionally signals might be detectable which were hidden in the noise for T450 (*e.g.* Si-Cl vibrations in IR vibrations). In order to doubtlessly prove the existence of tripodal site, it would be necessary to etect and distinguish the Cl-Ti stretch of a monopodal species from the asymmetric/symmetric stretching mode of a bipodal Cl_2-Ti site and monitor their intensity during thermal post-treatment.

A similar restructuring was attempted with other metal chlorides or alkoxides grafted in hexane, however an incomplete removal of the solvent after grafting resulted in the formation of carbon black on the sample surface. Due to this insufficient solvent removal and the interference of the residual carbon species characterization was complicated. However, using chemical vapor deposition with other metal precursors characterized by a low vapor pressure could be achieved by increasing the temperature or decreasing the vacuum during the precursor transfer step (see experimental figure 69, step 2). The latter would be the preferred option as it simultaneously decreases the concentration of moisture.

Further discoveries com from the comparison of the kinetic performance of TS-1 with T450. The behavior of the latter has been studied in epoxidation at temperatures between 40 and 80°C, in H_2O_2 decomposition as well as the influence of water on allyl alcohol epoxidation. Comparing the activity of TS-1 to T450 in allyl alcohol epoxidation under similar conditions could provide useful information about the active species in TS-1 as it was completed for the decomposition of hydrogen peroxide. The insight obtained on the active species of T450 by computational prediction of the rate determining oxygen transfer step (Figure 50) can be used to draw structure-activity relationships for TS-1. A series of cyclooctene epoxidation experiments with varying initial olefin concentration revealed a non-linear behavior of the initial rate constants (figure 67). The levelling off at higher cyclooctene concentrations indicates the participation of the olefin in the rate determining reaction step. Since it is assumed that the olefin is not coordinating to the Ti the oxygen transfer step is most likely the rate determining step in the epoxidation with hydrogen peroxide as with TBHP. It needs to be investigated whether a similar effect is observed for varying hydrogen peroxide concentrations while keeping the water content constant. Moreover, it would be of interest to see if the same non-linear behavior is observed at 80 and 50°C since the preferred reaction mechanism appears to be temperature dependent as shown in figure 57.

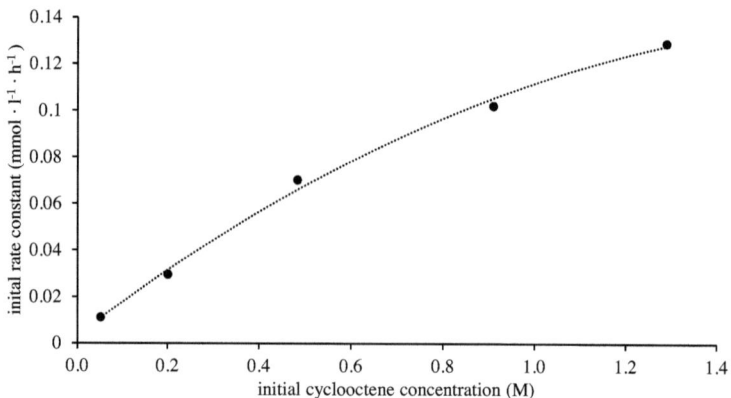

Figure 67: Initial rate constant of kinetic batch experiments with different initial cyclooctene concentrations; reaction conditions: 80 mg T450, 1 M H_2O_2 (30 wt% in water) in butanol at 65°C.

The performance of TS-1 is mostly assigned to the hydrophobic environment which the micropores create at the active sites[256,346]. The new insight into the active species of epoxidation with aqueous H_2O_2 indicates that water is not blocking the active sites as reported by Stöckmann et al.[345]. Water probably acts as a proton acceptor for the very acidic μ^2-hydroperoxo Ti species[217] which is responsible for oxidant decomposition as pointed out by kinetic experiments of this work. Following this approach, it might not be necessary to avoid water in the reaction mixtures but the presence of proton acceptors. Addition of small amounts of an inert Lewis acid might be sufficient in order to achieve pH values under which the μ^2-peroxo species is protonated.

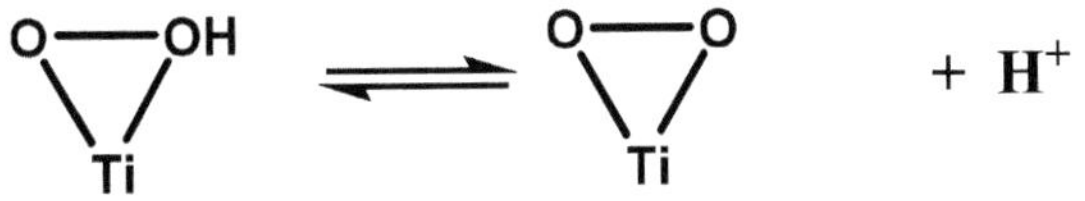

$$\underset{Ti}{O{-}OH} \rightleftharpoons \underset{Ti}{O{-}O} + H^+$$

Figure 68: Deprotonation of of μ^2-hydroperoxo-Ti surface species

Attention must be taken to avoid too acidic reaction solutions which can cause oxidant decomposition or epoxide opening. Circumventing the latter may be achievable through the use of very bulky lewis acids.

The same reason as above may cause the two different temperature ranges of T450 in epoxidation with hydrogen peroxide. At a certain temperature, proton dissociation may be entropically favored leading to the very active doubly deprotonated peroxo-Ti species. Due to this it would be necessary to decrease the temperature in order to avoid the second deprotonation. An increase in Ti-loading might be advantageous to maintain high conversions.

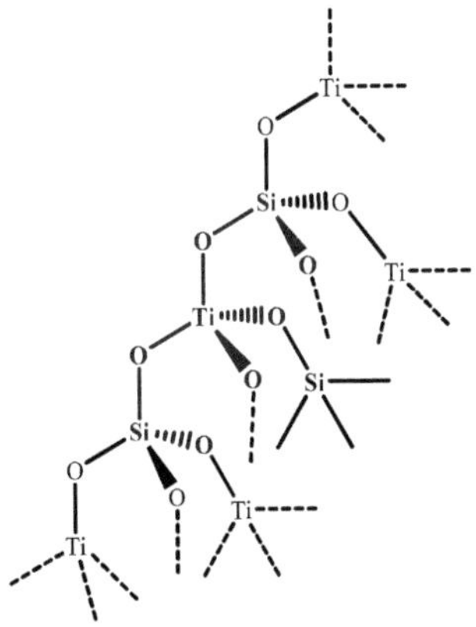

Figure 69: Titanium-silica network with the maximum loading of tetrahedrally coordinated Ti-atoms (smallest unit color in red)

Theoretically the most active atomic structure, exhibiting the highest Ti loading additionally to site isolation, can be achieved by alternated bonding of SiO_4- and TiO_4-tetrahedra having a maximum Ti-loading of 23.5 wt% (figure 68). This type of material cannot be produced by chemical vapor deposition or grafting in liquid phase unless it is deposited on a surface in a fashion of atomic layer deposition. However, the synthesis method of the unsupported material should yield a high surface area material and therefore lead to good accessibility of the Ti atoms. This leaves only flame pyrolysis or sol-gel procedure of $TiCl_4$ with $Si(O^iPr)_4$ as potential preparation methods. With flame pyrolysis the structurizing step is poorly controllable, as the high flame temperature most likely yields thermodynamically more stable oligomeric TiO_2-clusters at Ti concentration higher than 1 wt% Ti and not to the desired alternating tetrahedral structure[336]. In the sol-gel preparation the structuring step is easily controllable through the proper choice of chemicals, and the optimal loading could be increased to 10 wt% Ti in the epoxidation with cumyl hydroperoxide. However, at higher loadings than 10 wt% those aerosol catalysts were less active in cyclohexene epoxidation probably due to the formation of TiO_x cluster.

Part V

Appendix

8 Experimental Section

Material synthesis

Pretreatment of silica before grafting

Silica powder (Aerosil 200® from Degussa, specific surface area of 200 m$^2 \cdot$ g^{-1}) was impregnated with water and dried overnight at 120 °C in a vacuum oven resulting in a more convenient handling of the material. Subsequently, the material was dehydrated at 700 °C (T_{pre} = 700 °C) under 5 μbar dynamic vacuum and stored in a glove box (<1 ppm O$_2$ and H$_2$O) in order to avoid re-adsorption of water.

Preparation of T$_{dimer}$

T450 with a Ti loading of 0.8±0.1 wt% (*i.e.*, 0.18 mmol $\cdot$ g^{-1}) was used as starting point to synthesize a Ti-O-Ti containing material upon reaction of the Ti-Cl bonds with titanium(VI)isopropoxide. To this, 22 mL of dry hexane was added to 300 mg of Ti$_{mono}$ (containing 54 μmol Ti-Cl bonds in total) in a 50 mL round-bottom flask under a nitrogen atmosphere. After addition of 0.8 mL of pre-distilled Ti(O^{i}Pr)$_4$ (Aldrich; 2.7 mmol, *i.e.* 50× excess relative to the TiCl bonds), the solution was stirred for 7 hours. Subsequently, the solvent was removed by cold distillation. Finally, the material was dried under a dynamic vacuum of 10 μbar overnight.

Grafting of Ti(O-^{i}prop)$_4$ to silylated silica in hexane

This support was prepared by silylating Aerosil200 (preheated at 700 °C under a dynamic vacuum of 10 μbar) with N,N-bis-(trimethylsilyl)methylamine. IR spectroscopy showed the removal of all silanols and no evidence for a methyl group of the used silylation agent was detected. The latter is evidence for the complete removal of excess amine. 200 mg of silylated silica was mixed with 5 mL of dry hexane in a 50 mL round-bottom flask under a nitrogen atmosphere. After adding 0.5 mL of pure Ti(O^{i}Pr)$_4$, the solution was stirred for three hours. Subsequently, the solution was washed three times with 10 mL dry hexane. Finally the material was dried under a dynamic vacuum of 10 μbars overnight.

Grafting of TiCl$_4$ (T_{mono})

TiCl$_4$ (ACROS, three times distilled before use, colorless) was deposited onto less than 400 mg of material at $\approx$10 μbar static vacuum. A typical deposition (figure 69) consisted of a evacuation phase **(1)** (at pressures lower than 20 μbars), a transfer phase **(2)** (5-10 minutes), a reaction phase **(3)** (30 minutes), a desorption phase **(4)** under dynamic vacuum to evaporate excess TiCl$_4$ (5-10 minutes), and finally a post-grafting heat treatment at 50 °C $\leq$ T_{post} $\leq$ 450 °C (60 minutes).

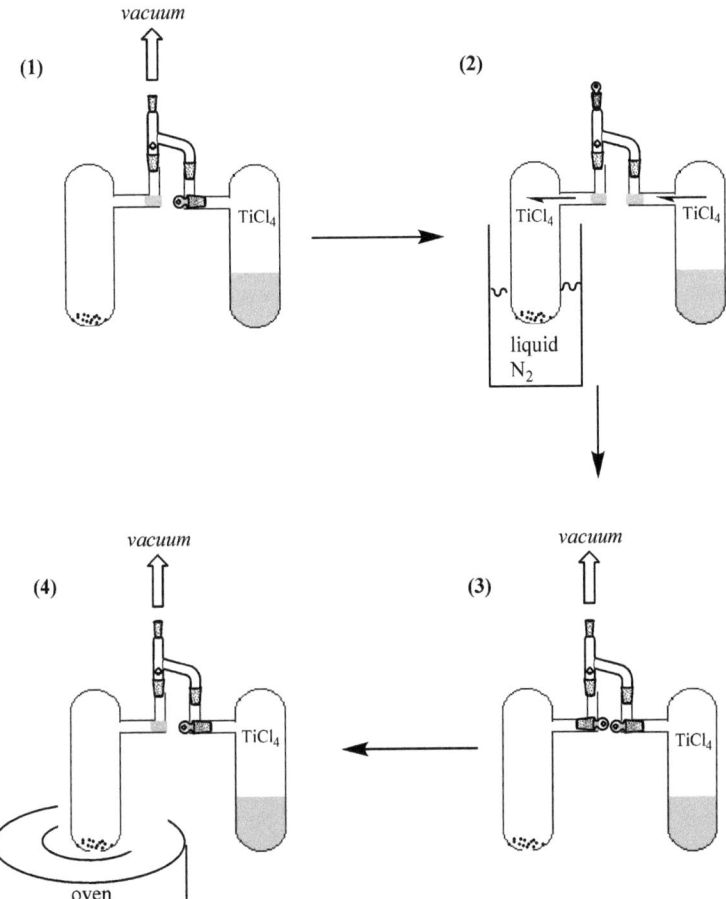

Figure 70: Sketch diagram of CVD procedure

Grafting of VOCl₃ and CrO₂Cl₂

Grafting of $VOCl_3$ (Strem Chemicals, distilled three times before use, slightly yellow color) or CrO_2Cl_2 (Aldrich, 99.99+% purity, distilled three times before use, orange color) to thermally dehydrated silica (Aerosil 200® from Degussa, specific surface area of 200 m² · g⁻¹) was performed as described above for $TiCl_4$. Materials are denoted, based on the metal and post-treatment temperature. *e.g.* Cr250 represents CrO_2Cl_2 grafted to silica pretreated at 700 °C, with a post-treatment at 250 °C under 10 μbar dynamic vacuum.

Grafting of Ti(O-ⁱprop)₄ in hexane

The Ti(O-ⁱprop)₄-grafted benchmark catalyst was synthesized as follows. 500 mg Aerosil200© preheated to 700°C was mixed into 5 mL of dry hexane in a 50 mL round-bottom flask under a nitrogen atmosphere. After adding 0.05 mL of pure titanium *iso*-propoxide, the solution was left stirring for three hours. Subsequently, the solution was washed three times with 10 mL dry hexane. Finally the Ti(O-ⁱprop)₄-catalyst was dried under a dynamic vacuum of 10 μbars overnight.

Spectroscopic characterization methods

^{1}H- and ^{13}C-MAS NMR spectroscopy

^{1}H- and ^{13}C-NMR spectra were acquired on an Avance NMR spectrometer (Bruker, Karlsruhe, Germany) operating at a ^{1}H Larmor frequency of 700 MHz. The samples were spun around the Magic Angle with a rate of 12 kHz at room temperature using a double resonance 4 mm probe (containing ±40 mg sample). The probe was tuned to the resonance frequencies of ^{1}H (700.13 MHz) and ^{13}C (176.06 MHz). The ppm scale of the spectra was calibrated using the ^{13}C signal of adamantane as an external secondary reference. In total three spectra were acquired from each samples. First, an ^{1}H spectrum was acquired using a single 90 degree pulse of 4.0 μs duration for excitation and a recycle delay of 2 s. Second a ^{13}C spectrum was acquired using a Cross Polarization (CP) sequence with TPPM decoupling of ^{1}H during acquisition. The initial 90 degree pulse was calibrated to 3.1 μs (81 kHz), the CP period with a duration of 1.5 ms employed a linear ramp from 70 % to 100 % and was experimentally optimized to give maximum intensity on ^{13}C. The decoupling during acquisition employed the same rf field amplitude as the 90 degree pulse (81 kHz). The TPPM had a phase difference of 15 degrees between pulses and a duration of 6.5 μs per pulse. The acquisition time was 29.1 ms with 4096 point and a sweep-width of 400 ppm. A total of 4096 scans were added with a recycle delay of 2 s. The FID's were processed by zero filling to 8k points and apodisation with a 128 Hz Lorentzian prior to Fourier Transform and phasing.

^{35}Cl-NMR spectroscopy

The ^{35}Cl-NMR spectra were acquired on a Varian/Chemagnetics Infinity+ spectrometer (Agilent, Palo Alto, USA) using a 6 mm DR MAS probe-head (containing ±100 mg sample). The probe was tuned to the resonance frequency of ^{35}Cl (around 49 MHz) and the sample was not rotated and was packed in custom made containers of PMMA with caps equipped with o-rings to ensure a gastight fit. The PMMA containers (instead of ZrO_2 rotors) proved crucial to avoid a background signal of the satellite transitions of ^{91}Zr. The samples were packed inside a glove box and during the experiments a continuous flow of $N_2(g)$ was maintained around the sample container. The spectra are skyline superpositions of a number of individual spectra with varying center frequencies. For each experiment the tuning and matching was optimized for the respective center frequency used. A QCPMG experiment was used to acquire the data according to the method of Schurko with identical WURST pulses for excitation and refocusing[283-284]. The WURST pulses used an exponent of 80 and a symmetric frequency sweep from -400 to +400 kHz. The pulses had a duration of 50 µs with a maximum amplitude of the rf field of approximately 22 kHz. In addition to the rf amplitude, the delays τ_1, τ_2 and τ_3 were carefully adjusted using a test sample of Cp_2TiCl_2 to values of 28.0, 3.0 and 10 µs. Per scan a total of 100 echoes were acquired. Each of these was digitized in 100 points with a dwell time of 1.0 µs, leading to a spectral width of 1.0 MHz and 10000 acquired points per QCPMG echo train. Processing of the data was done using Matlab (The MathWorks, Inc., Natick, MA, USA) in connection with the MatNMR toolbox. Each of the echoes was apodised with a sine squared function centered at the echo maximum (not for Cp_2TiCl_2 and $CpTiCl_3$ samples). The whole echo train was then multiplied with a Gaussian function with a width of 512 Hz, zero filled to 16k points and Fourier transformed. Finally the absolute mode was taken. The superposition of the spectra followed by a skyline projection produced the spectra as presented in the figures. Samples were independently measured twice to verify the reproducibility.

For the two test samples Cp_2TiCl_2 and $CpTiCl_3$, 1024 and 2048 scans were co-added with all other parameters identical (Figure ESI2). The amount of signal from the $CpTiCl_3$ sample was considerable less than from Cp_2TiCl_2. This was mainly caused by a much faster decay of the echo amplitude, indicating a faster transverse relaxation rate in this sample.

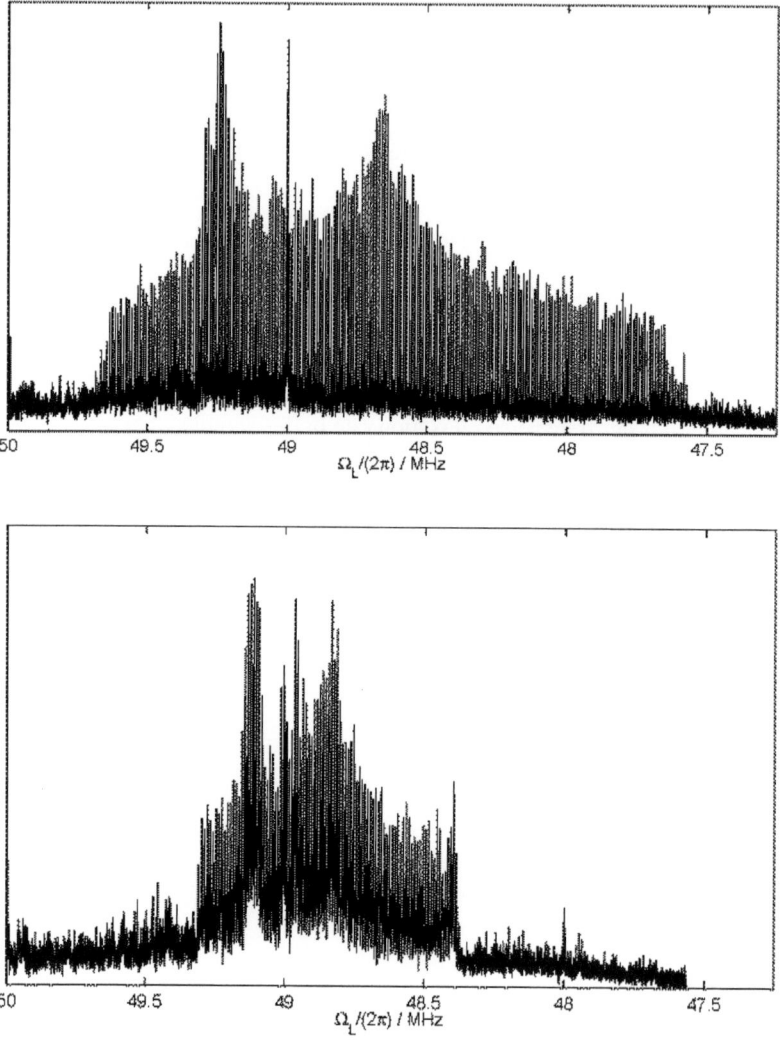

Figure 71. Left-hand-side: Cp_2TiCl_2 ; right hand-side: $CpTiCl_3$.

^{51}V-MAS NMR spectroscopy

. ^{51}V-NMR spectra were acquired on an Avance 400 NMR spectrometer (Bruker, Karlsruhe, Germany) operating at a ^{51}V Larmor frequency of 105 MHz. The samples were spun around the Magic Angle with a rate of 12.00 kHz at room temperature inside a 4 mm ZrO sample rotor (containing about 40 mg sample). The probe was tuned to the resonance frequency of ^{51}V (105.27 MHz) and 57344 scans were acquired. The experiment was a single pulse-aquire with a pulse length of 0.6 μs and a 0.5 s recycle delay. The acquisition time was 5.2 ms with 2048 points and a sweep-width of 1900.5 ppm. For the sample V450, 454976 scans were acquired in order to improve the S/N ratio.

IR spectroscopy

All IR measurements of the solid samples were completed in transmission mode, on a self-supporting wafer using a Bruker ALPHA R instrument inside a glove box. Time resolved IR measurements were conducted on a Bruker VERTEX 70v in a home-made glass cell with KBr windows.

UV-Vis spectroscopy

Diffuse Reflectance UV-Vis spectra were recorded with a Maya 2000 spectrometer (Ocean optics) equipped with a UV-Vis deuterium/halogen light source (DH-2000-BAL from Mikropack) against a $BaSO_4$ background.

X-ray Photon spectroscopy (by Susos)

XPS analysis was performed using a PhI5000 VersaProbe spectrometer (ULVAC-PHI, INC.) equipped with a 180° spherical capacitor energy analyzer and a multi-channel detection system with 16 channels. The samples were fixed on the sample holder insida a glove box and transferred into the XPS sample chamber in a glove bag avoiding any contact with moisture or air. Spectra were acquired at a base pressure of 5×10^{-8} Pa using a focused scanning monochromatic Al-K$_\alpha$ source (1486.6 eV) with a spot size of 200 μm. The instrument was run in the FAT (fixed analyzer transmission mode) mode with electrons emitted at 45° to the surface normal. Charge neutralization utilizing both a cool cathode electron flood source (1.2 eV) and very low energy Ar$^+$–ions (10 eV) were applied throughout the analysis.

XAS spectroscopy

XAS measurements at the Ti K-edge (E = 4,976 eV) were performed at the X10DA (SuperXAS) beamline at the Swiss Light Source (SLS), Villigen, Switzerland. The samples were packed and sealed in capton foil inside a glove box. Scans were performed with a Si(111) crystal monochromator, and were collected in fluorescence mode using a germanium detector from 4.900 to 5.800 keV. In the EXAFS analysis of the spectra recorded, multiple scans were averaged to improve the signal-to-noise ratio and fitted to actual XAS-measured data of Ti foil. The raw XAS data was background corrected, and normalized using the Athena software package. Shell fits were performed in R-space (1.1 < R < 3.0 Å or 1.1 < R < 3.4 Å for the dimer catalyst) after Fourier transformation (3 < k < 10 Å^{-1}) of k^3-weighted spectrum.

Raman spectroscopy

Raman measurements were performed with an in Via Raman microscope from Renishaw using a 325 nm (5 mW) of an IK series He-Cd laser. Sample were transferred in a nitrogen atmosphere to the RAMAN spectrometer and measured immediately after the samples were taken out of the transfer container.

Probing of the surface species

N,N-Bis(trimethylsilyl)methylamine (Sigma Aldrich, three times distilled before use) was contacted with the samples in the same way as TiCl$_4$, and heated from 50 to 250 °C at 5 μbars dynamic vacuum.

N,N-bis(trimethylsilyl)methylamine with a ^{13}C-labelled methyl-group was synthesized by adding 32 mL dry diethyl ether under an inert nitrogen atmosphere to 1.88 g lithium hexamethylsilizane, in a 2-neck-round-bottom-flask immersed in an ice bath. After 10 minutes, 0.65 mL of ^{13}C labeled methyl iodide were added. After 15 minutes the milky-yellowish liquid was heated up to reflux temperature (35 °C) meanwhile turning transparent. The mixture was left reacting for 48 h, and the liquid phase was transferred into another flask *via* cold distillation. Finally the solvent was evaporated under vacuum.

Bulk analysis

Ti and Cl were quantified with ICP-OES (Ultima 2 from Horiba Jobin Yvon), after digestion with HF or H_2SO_4 (both methods producing similar results). The Cl content was determined by precipitation of silver chloride (digestion with H_2SO_4). A known amount of catalyst was dissolved in a 200 ppm $AgNO_3$ solution containing 1 M H_2SO_4; all remaining solid material was centrifuged off after 24 h. The chlorine content of the catalyst was obtained from the difference of silver concentration before and after addition of the catalyst.

Catalytic experiments

Continuous flow experiments

Continuous epoxidation experiments were performed in a flow reactor (internal diameter 1.5 mm) with catalyst particles sieved to 200-250 µm to avoid back-mixing and a large pressure-drop. Products were quantified against a decane internal standard with GC (HP6890; HP-5 column, 30 m/0.32 mm/0.25 mm; flame ionization detector).

The catalytic performance of the Ti_{mono} and Ti_{dimer} materials was evaluated for the continuous flow epoxidation of several olefins using *tert*-butylhydroperoxide as limiting reactant (100 mM). The catalyst (mesh-size 150±50 µm) was densely packed in a fixed-bed reactor (Teflon® tube of 1.3 mm diameter) between quartz wool. No compacting of the column was observed. Samples were analyzed at various resident times and the products quantified against a decane internal standard with GC (HP6890; HP-5 column, 30 m/0.32 mm/0.25 mm; flame ionization detector).

The catalytic performance of the T450 materials were evaluated for the continuous flow epoxidation of several olefins using hydrogen peroxide as limiting reactant (250 mM). The catalyst (mesh-size 150±50 μm) was densely packed in a fixed-bed reactor (Teflon® tube of 1.3 mm diameter) between quartz wool. No compacting of the column was observed. Samples were analyzed at various resident times and the products quantified against a decane internal standard with GC (HP6890; HP-5 column, 30 m/0.32 mm/0.25 mm; flame ionization detector).

Batch experiments

Kinetic batch experiments with varying cyclooctene concentration in *n*-butanol were completed in a 2-neck-50mL-round-bottom-flask equipped with a magnetic stirrer, a septum and a condenser. The catalyst was first heated in the pure solvent with the desired amount of cyclooctene (0.2, 0.5, 1.0 or 1.5 mL) until the reaction temperature was reached and subsequently the hydrogen peroxide (30 wt%; 1 mL) or the *tert*-butylhydroperoxide (5.5 M in decane; 1.5mL) was added. Samples were analyzed at various resident times and the products quantified against a decane internal standard with GC (HP6890; HP-5 column, 30 m/0.32 mm/0.25 mm; flame ionization detector).

Hydrogen peroxide decomposition

H_2O_2 decomposition experiments were conducted in a 3-neck-round-bottom-50mL flask equipped with a magnetic stirrer, a condenser, a septum for sampling and a thermometer. The catalyst was heated in the flask to the reaction temperature and subsequently the reaction solution was added. The H_2O_2 concentration was determined by titrating a 1M Ce^{+3} diluted sulfuric acid solution to the reaction sample until the solution turned yellow. A calibration curve was determined before every experiment in order to avoid systematic errors due to aging of the Ce^{3+}-solution.

Computational methods

Quantum chemical calculations were performed with the Gaussian09 software package[292] at the UB3LYP/6-311++G(df,pd)//UB3LYP/6-31G(d,p) level of theory,[293] using a LANL2DZ basis set with an additional f-polarization function for Ti[294]. The reported relative energies of the stationary points on the Potential Energy Surfaces (PESs, *viz.* the energy barriers E_b and reaction energies $\Delta_r H$) were corrected for Zero-Point-Energy (ZPE) differences and are reported at 0 K (adiabatic PESs). Computed frequencies were scaled by a factor 0.9614[295]. The polarizable continuum model (PCM)[296]. as implemented in Gaussian09, was used to take into account solvent effects when necessary. The rigid, hydrogen-bonded tris(siloxide) assembly reflects the inherent constrains of such surface species, while allowing for some flexibility.

9 References

[1] a) B. P. C. Hereijgers, R. F. Parton, B. M. Weckhuysen *ACS Catal.* **2011**, *1*, 1183–1192; b) Vision Paper, 'Strategic Research Agenda and Implementation Action Plan of the European Technology Platform on Sustainable Chemistry', **2008**, available at http://www.suschem.org.

[2] F. Cavani, J. H. Teles, *ChemSusChem* **2009**, *2*, 508.

[3] I. Hermans, E. S. Spier, U. Neuenschwander, N. Turrà, A. Baiker, *Top. Catal.* **2009**, *52*, 1162.

[4] a) 'Metal-Catalyzed Oxidations of Organic Compounds', R. A. Sheldon, J. K. Kochi, Academic Press, Amsterdam, **1981**; b)'Thermochemical Kinetics', S. W. Benson, John Wiley & Sons, Inc., New York, **1968**; c) 'Catalytic Asymmetric Synthesis', R. A. Johnson, K. B. Sharpless, VCH Publishers, Inc., New York, **1993**; d),Encyclopedia of Catalysis', M. L. Merlau, C. C. Borg-Breen, S. T. Nguyen, John Wiley & Sons, Inc., New York, **2002**; e) Weitz, E., Scheffer, A. *Chem. Ber.* **1921**, *54*, 2327; f) D. Enders, J. Zhu, G. Raabe, *Angew. Chem., Int. Ed. Engl.* **1996**, *35*, 1725.

[5] I. Hermans, T. L. Nguyen, P. A. Jacobs, J. Peeters, *ChemPhysChem* **2005**, *6*, 637.

[6] I. Hermans, P. A. Jacobs, J. Peeters, *J. Mol. Catal. A: Chem.* **2006**, *251*, 221.

[7] L. Vereecken, T. L. Nguyen, I. Hermans, J. Peeters, *Chem. Phys. Lett.* **2004**, *393*, 432.

[8] I. Hermans, J. Peeters, L. Vereecken, P. Jacobs, *ChemPhysChem* **2007**, *8*, 2678.

[9] I. Hermans, J. Peeters, P. Jacobs, *J. Org. Chem.* **2007**, *72*, 3057.

[10] I. Hermans, P. A. Jacobs, J. Peeters, *Chem. Eur. J.* **2007**, *13*, 754.

[11] I. Hermans, P. A. Jacobs, J. Peeters, *Chem. Eur. J.* **2006**, *12*, 4229.

[12] a) U. Neuenschwander, F. Guignard, I. Hermans, *ChemSusChem* **2010**, *3*, 75; b) U. Neuenschwander, E. Meier, I. Hermans, *Che. Eur. J.* **2012**, *18*, 6776-6780.

[13] N. Turrà, U. Neuenschwander, A. Baiker, J. Peeters, I. Hermans, *Chem. Eur. J.* **2010**, *16*, 13226.

[16] I. Hermans, E. Breynaert, H. Poelman, R. De Gryse, D. Liang, G. Van Tendeloo, A. Maes, J. Peeters, P. Jacobs, *Phys. Chem. Chem. Phys.* **2007**, *9*, 5382.

[17] I. Hermans, P. A. Jacobs, J. Peeters, *ChemPhysChem* **2006**, *7*, 1142.

[19] I. Hermans, L. Vereecken, P. A. Jacobs, J. Peeters, *Chem. Comm.* **2004**, 1140.

[20] I. Hermans, P. Jacobs, J. Peeters, *Phys. Chem. Chem. Phys.* **2007**, *9*, 686.

[21] I. Hermans, P. A. Jacobs, J. Peeters, *Phys. Chem. Chem. Phys.* **2008**, *10*, 1125.

[22] I. Hermans, J. van Deun, K. Houthoofd, J. Peeters, P. Jacobs, *J. Catal.* **2007**, *251*, 204.

[23] C. Aellig, C. Girard, I. Hermans, *Angew. Chem. Int. Ed.* **2011**, *50*, 12355.

[24]   a) F. S. Bridson Jones, G. D. Buckley, L. H. Cross, A. P. Driver, *J. Chem. Soc.* **1951**, 2999; b) G. D. Buckley, F. S. Bridson Jones,W. J. Levy, D. C. Rogers, *Br. Pat. 668 309*, **1949**; c) G. D. Buckley, A. P. Driver, F. S. Bridson Jones, *Br. Pat. 649 680*, **1949**.

[25]   a) G. I. Panov, K. A. Dubkov, E. V. Starokon, V. N. Parmon, *React. Kinet. Catal. Lett.* **2002**, *76*, 401; b) G. I. Panov, K. A. Dubkov, E. V. Starokon, V. N. Parmon, *React. Kinet. Catal. Lett.* **2002**, *77*, 197; c) E. V. Starokon, K. A. Dubkov, D. E. Babushkin, V. N. Parmon, G. I. Panov, *Adv. Synth. Catal.* **2004**, *346*, 268; d) S. V. Semikolenov, K. A. Dubkov, E. V. Starokon, D. E. Babushkin, G. I. Panov, *Russ. Chem. Bull. Int. Ed.* **2005**, *54*, 948; e) E. V. Starokon, K. A. Dubkov, V. N. Parmon, G. I. Panov, *React. Kinet. Catal. Lett.* **2005**, *84*, 383.

[26]   I. Hermans, B. Moens, J. Peeters, P. A. Jacobs, B. Sels, *Phys. Chem. Chem. Phys.* **2007**, *9*, 4269.

[27]   I. Hermans, K. Janssen, B. Moens, A. Philippaerts, B. Van Berlo, J. Peeters, P. A. Jacobs, B. F. Sels, *Adv. Synth. Catal.* **2007**, *349*, 1604.

[28]   *Chem. Eng. News* **2006**, *84*, 30; b) http://www.basf.com/group/pressemitteilungen/ P-09-461.

[29]   a) D. Vandenberg, J.-P. Ganhy, N. Vanlautem (Solvay), WO9940024, **1999** b) A. A. Frimer, *Chem. Rev.* **1979**, *79*, 359; c) D. R. Kearns, *Chem. Rev.* **1971**, *71*, 395; d) E. L. Clennan, *Tetrahedron* **2000**, *56*, 9151.

[30]   a) M. Arab, D. Bougeard, J. M. Aubry, J. Marko, J. F. Paul, E. Payen, *J. Raman Spec.* **2002**, *33*, 390; b) J. M. Aubry, B. Cazin, *Inorg. Chem.* **1988**, *27*, 2013; c) J. M. Aubry, *J. Am. Chem. Soc.* **1985**, *107*, 5844; d) V. Nardello, S. Bouttemy, J. M. Aubry, *J. Mol. Catal.* **1997**, *117*, 439; e) V. Nardello, J. Marko, G. Vermeersch, J. M. Aubry, *Inorg. Chem.* **1998**, *37*, 5418; f) J. Wahlen, D. E. De Vos, P. A. Jacobs, V. Nardello, J.-M. Aubry, P. L. Alsters, *J. Catal.* **2007**, *249*, 15; g) B. F. Sels, D. E. De Vos, P. A. Jacobs, *J. Am. Chem. Soc.* **2007**, *129*, 6926; h) J. Wahlen, D. E. De Vos, M. H. Groothaert, V. Nardello, J.-M. Aubry, P. L. Alsters, P. A. Jacobs, *J. Am. Chem. Soc.* **2005**, *127*, 17166; i) J. Wahlen, D. E. De Vos, P. A. Jacobs, P. L. Alsters, *Adv. Synth. Catal.* **2004**, *346*, 152.

[31]   J. Wahlen, D. De Vos, W. Jary, P. Alsters, P. Jacobs, *Chem. Commun.* **2007**, 2333.

[32]   V. Duma, D. Hönicke, *J. Catal.* **2000**, *191*, 93.

[33]   Buijink, J. K. F.; Lange, J.-P.; Bos, A. N. R.; Horton, A. D.; Niele, F. G. M. *Mechanisms in Homogeneous and Heterogeneous Epoxidation Catalysis* (Ed. Oyama, S.T.), **2008**, Elsevier, Amsterdam, 355.

[34]   J. T. Scanlan, D. Swern, *JACS* **1940**, *62*, 2305; M. Mugdan, D. P. Young, *J. Chem. Soc.*, **1949**, 2988

[35]   E. G. E. Hawkins, *J. Chem. Soc.* **1950**, 2169.

[36]   a) R. A. Sheldon, B.Cornils, W. A.Herrmann, *Applied Homogeneous Catalysis with Organometallic Compounds: A Comprehensive Handbook in Two Volumes*, Vol. 1,

Wiley-VCH Verlag GmbH, Weinheim, Germany, **1996**, p. 411; b) R. A. Sheldon, J. K. Kochi, *Metal-Catalyzed Oxidations of Organic Compounds*, Academic Press, Inc., New York, **1981**, p. 424; R. c) A. Sheldon, R.Ugo, *Aspects of Homogeneous Catalysis*, Vol. 4, D. Reidel, Dordrecht, **1981**, p. 3; d) K. A. Jorgensen, *Chem. Rev.* **1989**, *89*, 431; e) T. Katsuki, in Beller, M. and Bolm, C. *Epoxidations, in Transition Metals for Organic Synthesis: Building Blocks and Fine Chemicals*, Wiley-VCH Verlag GmbH, Weinheim, Germany, **2008**, pp. 337–377.

[37]   U.S. Pat. 3,350,4223,350,422, J. Kollar (to Halcon International, Inc.), **1967**

[38]   M. Beller, M. Baerns eds., *Basic Principles in Applied Catalysis*, Springer Series in Chemical Physics, Berlin, **2004**, *75*, p. 392

[39]   H. Mimoun, I. S. d. Roch, and L. Sajus, *Tetrahedron* **1970**, *26*, 37.

[40]   H. Arakawa, Y. Moro-oka, A. Ozaki, *Bull. Chem. Soc. Jpn.* **1974**, *47*, 2958.

[41]   J. Sobeczak and J. J. Ziolkowski, *Inorg. Chim. Acta.* **1976**, *19*, 15.

[42]   D. D. Agarwal, *J. Mol. Catal.* **1988**, *44*, 65.

[43]   R. A. Sheldon, J. A. Van Doorn, *J. Catal.* **1973**, *31*, 427.

[44]   M. N. Sheng, J. G. Zajacek, *J. Org. Chem.* **1970**, *35*, 1839.

[45]   W. R. Thiel, M. Angstl, N. Hansen, *J. Mol. Catal. A: Chem.* **1995**, *103*, 5.

[46]   G. A. Tolstikov, V. P. Yurev, U. M. Dzhemilev, *Russ. Chem. Rev.* **1975**, *44*, 645

[47]   S.-i. Yamada, T. Masahiko, S. Terashima, *J. Am. Chem. Soc.* **1977**, *99*, 1988

[48]   S. Coleman-Kammula, E. T. Duim-Koolstra, *J. Organomet. Chem.* **1983**, 53

[49]   A. Arcorci, F. P. Ballisteri, G. A. Tomaselli, F. di Furia, G. Modena, *J. Org. Chem.* **1986**, *51*, 2374.

[50]   B. E. Rossiter, T. R. Verhoeven, K. B. Sharpless, *Tetrahedron Lett.* **1979**, *49*, 4733.

[51]   G. B. Payne and P. H. Williams, *J. Org. Chem.* **1959**, *24*, 54.

[52]   Y. Matoba, H. Inoue, J. Akagi, T. Okabayashi, Y. Ishii, M. Ogawa, *Synth. Commun.* **1984**, *14*, 865.

[53]   C. Venturello, E. Alneri, M. Ricci, *J. Org. Chem.* **1983**, *48*, 3831.

[54]   K. Sato, M. Aoki, M. Ogawa, T. Hashimoto, D. Panyella, R Noyori, *Bull. Chem. Soc. Jpn.* **1997**, *70*, 905.

[55]   a) H. Mimoun L. Saussine, E. Daire, M. Postel, J. Fischer, R. Weiss, *J. Am. Chem. Soc.* **1983**, *105*, 3101; b) H. Mimoun, M. Mignard, P. Brechot, L. Saussine, *J. Am. Chem. Soc.* **1986**, *108*, 3711.

[56]   D. D. Agarwal, P. Sangha, *Indian J. Chem.* **1996**, *35B*, 267; b) S. K. Gupta, D. D. Agarwal, D. Raina, *Indian J. Chem.* **1996**, *35A*, 995.

[57]   a) C. J. Chang, J. A. Labinger, H. B. Gray, *Inorg. Chem.* **1997**, 36, 5927; b) J. E. Lyons, *Tetrahedron Lett.* **1974**, *32*, 2737.

[58]   J. H. Hwang, M. M. Abu-Omar, *Tetrahedron Lett.* **1999**, *40*, 8313.

[59]   K. B. Sharpless, R. C. Michaelson, *J. Am. Chem. Soc.* **1973**, *95*, 6136.

[60]   H. B. Henbest and R. A. L. Wilson, J. Chem. Soc. 1958–1965 (1957).

[61]   B. E. Rossiter, T. R. Verhoeven, and K. B. Sharpless, *Tetrahedron Lett.* **1979**, *49*, 4733.

[62]   a) C. Bolm and T. Kuhn, *Synlett* **2000**, 899; b) N. Murase, Y. Hoshino, M. Oishi, H. Yamamoto, *J. Org. Chem.* **1999**, *64*, 338; c) Y. Hoshino, H. Yamamoto, *J. Am. Chem. Soc.* **2000**, *122*, 10452; d) Y. Hoshino, N. Murase, M. Oishi, H. Yamamoto, *Bull. Chem. Soc. Jpn.* **2000**, *73*, 1653.

[63]   T. Katsuki, K. B. Sharpless, *J. Am. Chem. Soc.*, **1980**, *102*, 5974.

[64]   A. Pfenninger, *Synthesis* **1986**, 89

[65]   Y. Gao, R. M. Hanson, J. M. Klunder, S. Y. Ko, H. Masamune, K. B. J. Sharpless, *J. Am. Chem. Soc.* **1987**, *109*, 5765.

[66]   J. G. Hill, B. E. Rossiter, K. B. J. Sharpless, *J. Org. Chem.* **1983**, *48*, 3607.

[67]   a) D. J. Morgans, Jr., K. B. Sharpless, S. G. Traynor, *J. Am. Chem Soc.* **1981**, *103*, 462; b) L. D.-L. Lu, R. A. Johnson, M. G. Finn, K. B. Sharpless, *J. Org. Chem.* **1984**, *49*, 728

[68]   V. S.,.Martin, S. S. Woodward, T. Katsuki, Y. Yamada, M. Ikeda, K. B. Sharpless, *J. Am. Chem. Soc.* **1981**, *103*, 6237

[69]   K. B. Sharpless, S. S. Woodward, M. G. Finn, *Pure Appl. Chem.* **1983**, *55*, 1823.

[70]   K. B. Sharpless, C. H. Behrens, T. Katsuki, A. W. M. Lee, V. S. Martin, M. Takatani, S. M. Viti, F. J. Walker, S. S. Woodward, *Pure Appl. Chem.* **1983**, *55*, 589.

[71]   T. Katsuki, A. W. M. Lee, P. Ma, V. S. Martin, s. Masamuine, K. B. Sharpless, D. Tuddenham, F. J. Walker, *J. Org. Chem.* **1982**, *47*, 1373.

[72]   H. Kigoshi, M. Ojika, Y. Shizuri, H. Niwa, K. Yamada, *Tetrahedron Lett.* **1982**, *23*, 5413

[73]   K. Mori, H Ueda, *Tetrahedron* **1981**, *37*, 2581.

[74]   K. Mori, T. Otsuka, *Tetrahedron* **1983**, *39*, 3267.

[75]   C. Meister, H.-D. Scharf, *Liebigs Ann. Chem.* **1983**, 913

[76] T. Katsuki, K. B. Sharpless, *Eur. Pat. Appl.* EP46033; C. A. **1982**, *97*, 38838

[77] W. Adam, A. Griesbexk, E. Staab, *Tetrahedron Lett.* **1986**, *27*, 2839.

[78] a) M. Bailey, I. E. Marko, W. D. Ollis, *Tetrahedron Lett.* **1991**, *32*, 2687; b) M. Bailey, I. Staton, P. R. Ashton, I. E. Markó, W. D. Ollis, *Tetrahedron: Asymmetry* **1991**, *2*, 495.

[79] a) Y. Sawada, K. Matsumoto, S. Kondo, H. Watanabe, T. Ozawa, K. Suzuki, B. Saito, T. Katsuki, *Angew. Chem. Int. Ed.* **2006**, *45*, 3478–3480; b) Y. Sawada, K. Matsumoto, T. Katsuki, *Angew. Chem. Int. Ed.* **2007**, *46*, 24, 4559-4561; c) *Angew. Chem. Int. Ed.* **2009**, *48*, 7432-7435; d) S. Kondo, K. Saruhashi, K. Seki, K. Matsubara, K. Miyaji, T. Kubo, K. Matsumoto, T. Katsuki, *Angew. Chem. Int. Ed.* **2008**, *47*, 10195-10198; e) H. Egami, K. Matsumoto, T. Oguma, T. Katsuki, *J. Am. Chem. Soc.* **2010**, *132*, 5886–5895; f) A. Berkessel, M. Brandenburg, E. Leitterstorf, J. Frey, J. Lex, M. Schäfer, *Adv. Synth. Catal.* **2007**, *349*, 2385-2391; g) A. Berkessel, M. Brandenburg, M. Schäfer, *Adv. Synth. Catal.* **2008**, *350*, 1287-1294.

[80] a) J. H. Espenson, *J. Chem. Soc., Chem. Commun.* **1999** 479; b) G. S. Owens, J. Arias, M. M. Abu-Omar, *Catal. Today* **2000**, *55*, 317; c) W. A. Herrmann, F. E. Kuhn, *Acc. Chem. Res.* **1997**, *30*, 169; d) W. A. Herrmann, *J. Organomet. Chem.* **1995** *500*, 149; e) W. A. Herrman, R. W. Fischer, M. U. Rauch, W. Scherer, *J. Mol. Catal.* **1994**, *86*, 243.

[81] U.S. Patent 51663725166372 G. L. Crocco, R. J. Saxton, J. G. Zajacek, K. S. Wijesekera (to Arco Chemical Technology, L.P.) **1992**; DE 39023573902357 W. A. Herrmann, D. W. Marz, J. G. Kuchler, G. Weichselbaumer, R. W. Fischer, (to Hoechst AG), **1990**

[92] a) W. A. Herrmann, F. E. Kuhn, R. W. Fischer, W. R. Thiel, C. C. Romao, *Inorg. Chem.* **1992**, *31*, 4431; b) W. A. Herrmann, R. W. Fischer, D. W. Marz, *Angew. Chem., Int. Ed. Engl.* **1991**, *30*, 1638.

[83] a) A. Adolfsson, J. P. Chiang, C. Copéret, A. Yudin *J. Org. Chem.*, **2000**, *65*, 8651; b) W.-D. Wang and J. H. Espenson, *J. Am. Chem. Soc.* **1998**, *120*, 11335; c) F. E. Kühn, A. Scherbaum, W. A. Herrmann, *J. Organomet. Chem.* **1998**, *555*, 293.

[84] a) W. Adam, C. M. Mitchell, *Angew. Chem., Int. Ed. Engl.* **1996**, *35*, 533; b) A. K. Yudin, K. B. Sharpless, *J. Am. Chem. Soc.* **1997**, *119*, 11536; c) A. K. Yudin, J. P. Chiang, H. Adolfsson, C. Copéret, *J. Org. Chem.* **2001**, *66*, 4713

[85] H. Tan, J. H. Espenson, *Inorg. Chem.* **1998**, *37*, 467.

[86] C. Coperet, H. Adolfsson, K. B. Sharpless, *J. Chem. Soc., Chem. Commun.* **1997**, 1565

[87] W. Adam, C. M. Mitchell, C. R. Saha-Moller, *J. Org. Chem.* **1999**, *64*, 3699.

[88] (a) W. A. Herrmann, R. W. Fischer, W. Scherer, M. U. Rauch, *Angew. Chem. Int. Ed. Engl.* **1993**, *32*, 1157; (b) W. A. Herrmann, J. D. G. Correia, G. R. J. Artus, R. W.

Fischer, C.C. Romão, *J. Organomet. Chem.* **1996**, *520*, 139; (c) H. S. Glenn, K. A. Lawler, R. Hoffmann, W. A. Herrmann, W. Scherer, R. W. Fischer, *J. Am. Chem. Soc.* **1995**, *117*, 3231.

[89] P. Ferreira, W.M. Xue, É. Bencze, E. Herdtweck, F. E. Kühn, *Inorg. Chem.* **2001**, *40*, 5834.

[90] M. M. Abu-Omar, P. J. Hansen, J. H. Espenson, *J. Am. Chem. Soc.* **1996**, *118*, 4966.

[91] M. D. Skowronska-Ptasinska, M. L. W. Vortenbosch, A. A. van Santen, H. C. L. Abbenhuis, *Angew. Chem. Int. Ed.* **2002**, *41*, 637.

[92] a) Z. Gross, S. Nimri, *J. Am. Chem. Soc.* **1995**, *117*, 8021; b) J. T. Groves, J. Lee, S. S. Marla, *J. Am. Chem. Soc.* **1997**, *119*, 6269; c) J. P. Collman, A. S. Chien, T. A. Eberspacher, J. I. Brauman, *J. Am. Chem. Soc.* **2000**, *122*, 11098.

[93] a) J. T. Groves, T. E. Nemo, R. S. Myers, *J. Am. Chem. Soc.* **1979**, *101*, 1032; b) C. K. Chang, M.-S. Kuo, *J. Am. Chem. Soc.* **1979**, *101*, 3413; c) J. T. Groves, T. E. Nemo, *J. Am. Chem. Soc.* **1983**, *105*, 5786; d) J. T. Groves, T. E. Nemo, R. S. Myers, *J. Am. Chem. Soc.* **1979**, *101*, 1032; e) C. K. Chang, M.-S. Kuo, *J. Am. Chem. Soc.* **1979** *101*, 3413.

[94] a) S. Banfi, A. Maiocchi, A. Moggi, F. Montanari, S. Quici, *J. Chem. Soc., Chem. Commun.* **1990**, 1794; b) S. Takagi, B Takahashi, T. K. Miyamoto, Y. Sasaki, *Chem. Lett.* **1986**, 1275; c) S. Banfi, F. Montanari, S. Quici, *J. Org. Chem.* **1989**, *54*, 1850; d) F. Montanari, S. Banfi, S. Quici, *Pure Appl. Chem.* **1989**, *61*, 1631; e) B. Meunier, M. E. De Carvalho, A. Robert, *J. Mol. Catal.* **1987**, *41*, 185.

[95] a) A. Robert, B. Meunier, *New J. Chem.* **1988**, *12*, 885; b) T.-C. Zheng, D. E. Richardson, *Tetrahedron Lett.* **1995**, *36*, 833; c) B. De Poorter, B. Meunier, *Nouv. J. Chem.* **1985**, *9*, 393.

[96] a) J. Haber, R. Iwanejko, J. Poltowicz, P. Battioni, D. Mansuy, *J. Mol. Catal. A: Chem.* **2000**, *152*, 117; b) T.-C. Zheng, D. E. Richardson, *Tetrahedron Lett.* **1995**, *36*, 837; c) R. Ramasseul, C. Scheer, M. Tavares, J. C. Marchon, *J. Mol. Catal.* **1990**, *63*, 167; d) C. Querci, M. Ricci, *J. Chem. Soc., Chem. Commun.* **1989**, 889.

[97] a) D. Mohajer, S. Tangestaninejad, *J. Chem. Soc., Chem. Commun.* **1993**, 240–241; b) D. Mohajer, S. Tangestaninejad, *Tetrahedron Lett.* **1994**, *35*, 945.

[98] a) F. Waller, A. J. Bailey, W. P. Griffith, S. P. Marsden, E. H. Smith, *J. Mol. Catal. A: Chem.* **2000**, *154*, 85; b) S. Campestrini, A. Robert, B. Meunier, *J. Org. Chem.* **1991**, *56*, 3725.

[99] L. C. Yuan, T. C. Bruice, *J. Chem. Soc., Chem. Commun.* **1985**, 868.

[100] a) S. J. Yang, W. Nam, *Inorg. Chem.* **1998**, *37*, 606; b) G. X. He, T. C. Bruice, *J. Am. Chem. Soc.* **1991**, *113*, 2747.

[101] a) K. Machii, Y. Watanabe, I. Morishima, *J. Am. Chem. Soc.* **1995**, *117*, 6691; b) Y. Watanabe, K. Yamaguchi, I. Morishima, K. Takehira, M. Shimizu, T. Hayakawa,. H. Orita, *Inorg. Chem.* **1991**, *30*, 2581.

[102] F. Montanari, S. Banfi, S. Quici, *Pure Appl. Chem.* **1989**, *61*, 1631.

[103] T. Mori, Z. X. Jin, H. Tokuda, H. Nishino, T. Yoshida, *Chem. Pharm. Bull.* **1993**, *41*, 292.

[104] a) P. Battioni, J. F. Bartoli, P. Leduc, M. Fontecave, D. Mansuy, *J. Chem. Soc., Chem. Commun.* **1987**, 791; b) Y. Tsuda, K. Takahashi, T. Yamaguchi, S. Matsui, T. Komura, *J. Mol. Catal. A: Chem.* **1998**, *130*, 285.

[105] a) D. Mansuy, M. Fontecave, J. F. Bartoli, *J. Chem. Soc., Chem. Commun.* **1983**, 253; b) M. Fontecave D. Mansuy, *Tetrahedron* **1984**, *40*, 4297; c) T. Takai, E. Hata, T. Yamada, T. Mukaiyama, *Bull. Chem. Soc. Jpn.* **1991**, *64*, 2513; d) C. Bousquet, D. G. Gilheany, *Tetrahedron Lett.* **1995**, *36*, 7739.

[106] a) M. J. Gunter, P. Turner, *Coord. Chem. Rev.* **1991**, *108*, 115; b) B. Meunier, *Chem. Rev.* **1992**, *92*, 1411.

[107] D. Dolphin, T. G. Traylor, L. Y. Xie, *Acc. Chem. Res.* **1997**, *30*, 251.

[108] a) Y. Naruta, K. Maruyama, *Tetrahedron Lett.* **1987**, *28*, 4553; b) J. P. Collman, X. Zhang, R. T. Hembre, J. I. Brauman, *J. Am. Chem. Soc.* **1990**, *112*, 5356.

[109] M. L. Merlau, W. J. Grande, S. T. Nguyen, and J. T. Hupp, *J. Mol. Catal. A: Chem.* **2000**, *156*, 79.

[110] M. L. Merlau, M. P. Mejia, S. T. Nguyen, J. T. Hupp, *Angew. Chem., Int. Ed. Engl.* **2001**, *40*, 4239.

[111] B. Boitrell, V. Baveux-Chambenoît, P. Richard, *Helv. Chim. Acta*, **2004**, *87*, 2447

[112] Y. Naruta, F. Tani, N. Ishihara, K. Maruyama, *J. Am. Chem. Soc.* **1991**, *113*, 6865

[113] S. O'Malley, T. Kodadek, *J. Am. Chem. Soc.* **1989**, *111*, 9116.

[114] a) A. M. D. R. Gonsalves, M. M. Pereira, *J. Mol. Catal. A: Chem.* **1996**, *113*, 209; b) E. Guilmet, B. Meunier, *Nouv. J. Chim.* **1982**, *6*, 511; c) L.-C. Yuan, T. C. Bruice, *J. Am. Chem. Soc.* **1986**, *108*, 1643; d) B. Meunier, M. E. De Carvalho, O. Bortolini, M. Momenteau, *Inorg. Chem.* **1988**, *27*, 161.

[115] W. Zhang, J. L. Loebach, S. R. Wilson, E.N. Jacobsen, *J. Am. Chem. Soc.* **1990**, *112*, 2801.

[116] R. Irie, K. Noda, Y. Ito, N. Matsumoto, T. Katsuki, *Tetrahedron Lett.* **1990**, *31*, 7345.

[117] (a) E. N. Jacobsen in *Catalytic Asymmetric Synthesis* (Ed.: I. Ojima), VCH, New York, **1993**, p. 159; (b) T. Flessner, S. Doye, *J. Prakt. Chem.* **1999**, *341*, 436;

[118] (a) N. Hosoya, A. Hatayama, R. Irie, H. Sasaki, T. Katsuki, *Tetrahedron* **1994**, *50*, 4311; (b) H. Sasaki, R. Irie, T. Hamada, K. Suzuki, T. Katsuki, *Tetrahedron* **1994**, *50*, 11827.

[119] (a) T. Katsuki, *Adv. Synth. Catal.* **2002**, *344*, 131; (b) T. Katsuki, *Synlett* **2003**, 281; c) H. Nishikori, C. Ohta, T. Katsuki, *Synlett* **2000**, 1557.

[120] M. Palucki, N. Finny, P. J. Pospisill, M. J. Guler, T. Ishida, E. N. Jacobsen, *J. Am. Chem. Soc.* **1998**, *120*, 948.

[121] C. Linde, N. Koliaï, P.-O. Norrby, B. Åkermark, *Chem. Eur. J.* **2002**, *8*, 2568.

[122] a) C. Linde, N. Koliai, P.-O. Norrby, B. Åkermark, *Chem. Eur. J.* **2002**, *8*, 2568; b) W. Adam, K. J. Roschmann, C. R. Saha-Möller, D. Seebach, *J. Am. Chem. Soc.* **2002**, *124*, 5068.

[123] a) T. Mukaiyama, *Aldrichimica Acta* **1996**, *29*, 59; b) T. Mukaiyama, T. Yamada, *Bull. Chem. Soc. Jpn.* **1995**, *68*, 17; c) T. Yamada, K. Imagawa, T. Mukaiyama, *Chem. Lett.* **1992**, 2109.

[124] a) T. Mukaiyama, T. Yamada, T. Nagata, K. Imagawa, *Chem. Lett.* **1993**, 327; b) T. Nagata, K. Imagawa, T. Yamada, T. Mukaiyama, *Chem. Lett.* **1994**, 1259.

[125] S. Chang, N. H. Lee, E. N. Jacobsen, J. Org. Chem. 1993, 58, 6939.

[126] a) T. Nagata, K. Imagawa, T. Yamada, T. Mukaiyama, Bull. Chem. Soc. Jpn. 1995 68, 1455; b) T. Nagata, K. Imagawa, T. Yamada, T. Mukaiyama, Inorg. Chim. Acta 220, 283–287 (1994).

[127] R. Hage, J. E. Iburg, J. Kerschner, J. H. Koek, E. L. M. Lempers, R. J.. Martens, U. S. Racherla, S. W. Russell, T. Swarthoff, M. R. P. van Vliet,. J. B. Warnaar, L. van der Wolf and B. Krijnen, *Nature*, **1994**, *369*, 637.

[128] D. E. De Vos and T. Bein, *J. Organomet. Chem.* **1996**, *520*, 195.

[129] D. E. De Vos, Sels B. F., Reynaers M., Subba Rao Y. V., Jacobs P. A., *Tetrahedron Lett.* **1998**, *39*, 3221.

[130] A. Berkessel, C. A. Sklorz, *Tetrahedron Lett.* **1999**, *40*, 7965.

[131] R. W. Murray, *Chem. Rev.* **1989**, *89*, 1187.

[132] a) R. Mello, M. Fiorentino, O. Sciacovelli, R. Curci, *J. Org. Chem.* **1988**, *53*, 3890; b) D. Yang, X.-C. Wang, M.-K. Wong, Y.-C. Yip, M.-W. Tang, J. Am. Chem. Soc. 1996, 118, 11311.

[133] a) S. E. Denmark, D. C. Forbes, D. S. Hays, J. S. DePue and. R. G. Wilde, J. Org. Chem. 60, 1391–1407 (1995); b) S. E. Denmark, Z. Wu, C. Crudden, H. Matsuhashi, J. Org. Chem. 62, 8288–8289 (1997); c) S. E. Denmark, Z. Wu, J. Org. Chem. 1997, 62, 8964; d) S. E. Denmark, Z. Wu, J. Org. Chem. 1998, 63, 2810.

[134] M. Kurihara, S. Ito, N. Tsutsumi, N. Miyata, *Tetrahedron Lett.* **1994**, *35*, 1577.

[135] L. Shu, Y. Shi, *J. Org. Chem.* **2000**, *65*, 8807.

[136] a) Z.-X. Wang, Y. Tu, M. Frohn, J. R. Zhang, Y. Shi, *J. Am. Chem. Soc.* **1997**, *119*, 11224; b) Z.-X. Wang, S. M. Miller, O. P. Anderson, Y. J. Shi, *J. Org. Chem.* **2001**, *66*, 521.

[137] H. Wynberg, B. Marsman, *J. Org. Chem.* **1980**, *45*, 158.

[138] a) L. Alcaraz, G. Macdonald, J. Ragot, N. J. Lewis, R. J. K. Taylor, *Tetrahedron* **1999** *55*, 3707; b) G. Macdonald, L. Alcaraz, N. J. Lewis, R. J. K. Taylor, *Tetrahedron Lett.* **1998**, *39*, 5433; c) L. Alcaraz, G. Macdonald, J. Ragot, N. J. Lewis, R. J. K. Taylor, *J. Org. Chem.* **1998**, *63*, 3526.

[139] a) V. K. Aggarwal, M. F. Wang, *J. Chem. Soc., Chem. Commun.* **1996** 191b) M. F. A. Adamo, V. K. Aggarwal, M. A. Sage, *J. Am. Chem. Soc.* **2000**, *122*, 8317.

[140] A. Nelson, *Angew. Chem., Int. Ed. Engl.* **1999**, *38*, 1583.

[141] O. Bortolini, F. DiFuria, G. Modena, R. Seraglia, *J. Org. Chem.* **1985**, *50*, 2688.

[142] A. W. Johnson, R. B. LaCount, *J. Am. Chem. Soc.* **1961**, *83*, 417.

[143] a) A.-H. Li, L.-X. Dai, V. K. Aggarwal, *Chem. Rev.* **1997**, *97*, 2341; b) V. K. Aggarwal, J. G. Ford, A. Thompson, R. V. H. Jones, M. C. H. Standen, *J. Am. Chem. Soc.* **1996**, *118*, 7004.

[144] a) J. Wahlen, D. De Vos, P. A. Jacobs, *Org. Lett.* **2003**, *5*, 1777-1780; b) A. Berkessel, J. A. Adrio, *Adv. Synth. Catal.* **2004**, *346*, 275-280; c) A. Berkessel, J. A. Adrio, *J. Am. Chem. Soc.* **2006**, *128*, 13413-13419; d) A. Berkessel, *Angew. Chem. Int. Ed.* **2008**, *47*, 3677-3679; e) A. Berkessel, J. Krämer, F. Mummy, J.-M. Neudörfl, R. Haag, *Angew. Chem. Int. Ed.* **2013**, *52*, 739-743.

[145] G. A. Barf, R. A. Sheldon, *J. Mol. Catal. A: Chem.* **1995**, *102*, 23.

[146] G. Balavoine, C. Eskenazi, F. Meunier, H. Riviere, *Tetrahedron Lett.* **1984**, *25*, 3187.

[147] J. T. Groves, R. Quinn, *J. Am. Chem. Soc.* **1985**, *107*, 5790.

[148] E. G. Samsel, K. Srinivasan, J. K. Kochi, *J. Am. Chem. Soc.* **1985**, *107*, 7606.

[149] a) T. Takai, E. Hata, K. Yorozu, T. Mukaiyama, *Chem. Lett.* **1992**, 2077; b) R. I. Kureshy, N. H. Khan, S. H. R. Abdi, A. K. Bhatt, P. Iyer, *J. Mol. Catal. A: Chem.* **1997**, *121*, 25.

[150] a) H. Yoon, C. J. Burrows, *J. Am. Chem. Soc.* **1988**, *110*, 4087; b) G. Rousselet, C. Chassagnard, P. Capdevielle, M. Maumy, *Tetrahedron Lett.* **1996**, *37*, 8497.

[151] a) M. D. Jones, R. Raja, J. M. Thomas, B. F. G. Johnson, D. W. Lewis, J. Rouzaud, K. D. M. Harris, *Angew. Chem. Int. Ed.* **2003**, *42*, 4326; b) R. Raja, J. M. Thomas, B. F. G.

Johnson, D. E. W. Vaughan, *J. Am. Chem. Soc.* **2003**, *125*, 14982; c) R. Schögl, S. B. Abd Hamid, *Angew. Chem. Int. Ed.* **2004**, *43*, 1628–1637; d) J. Grunes, J. Zhu, G. A. Somorjai, *Chem. Commun.* **2003**, 2257–2260.

[152] a) J. H. Clark, *Supported Reagents in Organic Synthesis*, Wiley- VCH, Weinheim, **1994**; b) P. McMorn, G. J. Hutchings, *Chem. Soc. Rev.* **2004**, *33*, 108–122.

[153] S. Shylesh, M. Jia, W. R. Thiel, *Microreview*, **2010**, 4395-4410.

[154] a) D. Brunel, *Microporous Mesoporous Mater.* **1999**, *27*, 329–344; b) P. Barbaro, *Chem. Eur. J.* **2006**, *12*, 5666.

[155] a) A. Sayari, *Chem. Mater.* **1996**, *8*, 1840–1852; b) A. Sayari, S. Hamoudi, *Chem. Mater.* **2001**, *13*, 3151–3168; c) A. Taguchi, F. Schuth, *Microporous Mesoporous Mater.* **2004**, *77*, 1–45; d) J. M. Thomas, R. Raja, G. Sankar, R. G. Bell, *Acc. Chem. Res.* **2001**, *34*, 191–200.

[156] a) M. M. Miller, D. C. Sherrington, *J. Catal.* **1995**, *152*, 368– 376; b) M. M. Miller, D. C. Sherrington, *J. Catal.* **1995**, *152*, 377–383; c) M. M. Miller, D. C. Sherrington, *J. Chem. Soc.,Chem. Commun.* **1994**, 55–56; d) J. H. Ahn, D. C. Sherrington, *Chem. Commun.* **1996**, 643–644; e) D. C. Sherrington, *Catal. Today* **2000**, *57*, 87–104; f) M. M. Miller, D. C. Sherrington, S. Simpson, *J. Chem. Soc., Perkin Trans. 2* **1994**, 2091– 2096; g) G. Gelbard, F. Breton, M. Quenard, D. C. Sherrington, *J. Mol. Catal. A: Chem.* **2000**, *153*, 7–18.

[157] S. V. Kotov, S. Boneva, *J. Mol. Catal. A: Chem.* **1999**, *139*, 271–283.

[158] T. Szymanska-Buzar, J. J. Ziólkowski, *J. Mol. Catal.* **1981**, *11*, 371-381.

[159] a) T. Maschmeyer, F. Rey, G. Sankar, J. M. Thomas, *Nature* **1995**, *378*, 159–162; b) J. M. Thomas, R. Raja, D. W. Lewis, *Angew. Chem. Int. Ed.* **2005**, *44*, 6456–6482.

[160] a) P. Ferreira, I. S. Gonçalves, F. E. Kühn, A. D. Lopes, M. A. Martins, M. Pillinger, A. Pina, J. Rocha, C. C. Romão, A. M. Santos, T. M. Santos, A. A. Valente, *Eur. J. Inorg. Chem.* **2000**, 2263–2270; b) C. D. Nunes, A. A. Valente, M. Pillinger, J. Rocha, I. S. Gonçalves, *Chem. Eur. J.* **2003**, *9*, 4380–4390; c) B. Monteiro, S. S. Balula, S. Gago, C. Grosso, S. Figueiredo, A. D. Lopes, A. A. Valente, M. Pillinger, J. P. Lourenco, I. S. Gonçalves, *J. Mol. Catal. A* **2009**, *297*, 110–117.

[161] Q. Yang, C. Copéret, C. Li, J.-M. Basset, *New J. Chem.* **2003**, *27*, 319–323.

[162] P. C. Bakala, E. Briot, L. Salles, J.-M. Bregeault, *Appl. Catal. A: General* **2006**, *300*, 91–99.

[163] a) C. D. Nunes, A. A. Valente, M. Pillinger, A. C. Fernandes, C. C. Romão, J. Rocha, I. S. Gonçalves, *J. Mater. Chem.* **2002**, *12*, 1735–1742; b) C. D. Nunes, M. Pillinger, A. A. Valente, J. Rocha, A. D. Lopes, I. S. Gonçalves, *Eur. J. Inorg. Chem.* **2003**, 3870–3877.

[164] a) A. Sakthivel, J. Zhao, M. Hanzlik, F. E. Kühn, *Dalton Trans.* **2004**, 3338–3341; b) A. Sakthivel, J. Zhao, F. E. Kühn, *Catal. Lett.* **2005**, *102*, 115–199.

[165] A. Sakthivel, J. Zhao, M. Hanzlik, A. S. T. Chiang, W. A. Herrmann, F. E. Kühn, *Adv. Synth. Catal.* **2005**, *347*, 473–483.

[166] S. K. Maiti, S. Dinda, M. Nandi, A. Bhaumik, R. Bhattacharya, *J. Mol. Catal. A* **2008**, *287*, 135–141.

[167] a) M. Masteri-Farahani, F. Farzaneh, M. Ghandi, *J. Mol. Catal. A* **2006**, *243*, 170–175; b) M. Masteri-Farahani, *J. Mol. Catal. A* **2010**, *316*, 45–51.

[168] a) A. Hroch, G. Gemmecker, W. R. Thiel, *Eur. J. Inorg. Chem.* **2000**, 1107–1114; b) W. R. Thiel, *Chem. Ber.* **1996**, *129*, 575– 581; c) W. R. Thiel, *J. Mol. Catal. A* **1997**, *117*, 449–454.

[169] M. Jia, A. Seifert, M. Berger, H. Giegengack, S. Schulze, W. R. Thiel, *Chem. Mater.* **2004**, *16*, 877–882.

[170] S. Shylesh, J. Schweitzer, S. Demeshko, V. Schünemann, S. Ernst, W. R. Thiel, *Adv. Synth. Catal.* **2009**, *351*, 1789–1795.

[171] J. Jarupatrakorn, M. P. Coles, T. D. Tilley, *Chem. Mater.* **2005**, *17*, 1818–1828.

[172] a) G. J. Jin, G. Lu, Y. Guo, Y. Guo, J. Wang, X. Liu, *Catal. Lett.* **2003**, *87*, 249; b) G. Jin, G. Lu, Y. Guo, Y. Guo, J. Wang, X. Liu, W. Kong, X. Liu, *Catal. Lett.* **2004**, *97*, 191; c) G. Jin, G. Lu, Y. Guo, Y. Guo, J. Wang, W. Kong, X. Liu, *J. Mol. Catal. A* **2005**, *232*, 165.

[173] Z. X. Song, N. Mimura, J. J. Bravo-Suarez, T. Akita, S. Tsubota, S. T. Oyama, *Appl. Catal. A* **2007**, *316*, 142.

[174] C. Hammond, J. Straus, M. Righettoni, S. E. Pratsinis, I. Hermans, *ACS Catalysis*, **2013**, *3*, 321-327

[175] B. F. Sels, D. E. De Vos, P. A. Jacobs, *Tetrahedron Lett.* **1996**, *37*, 8557–8560.

[176] D. Hoegaerts, B. F. Sels, D. E. De Vos, F. Verpoort, P. A. Jacobs, Catal. Today 2000, 60, 209–218.

[177] A. L. Villa de P., B. F. Sels, D. E. De Vos, P. A. Jacobs, J. Org. Chem. 1999, 64, 7267–7270.

[178] a) K. Yamaguchi, C. Yoshida, S. Uchida, N. Mizuno, *J. Am. Chem. Soc.* **2005**, *127*, 530–531; b) J. Kasai, Y. Nakagawa, S. Uchida, K. Yamaguchi, N. Mizuno, *Chem. Eur. J.* **2006**, *12*, 4176–4184.

[179] S. Tangestaninejad, M. H. Habibi, V. Mirkhani, M. Moghadam, G. Grivani, *J. Mol. Catal. A* **2006**, *255*, 249–253.

[180] R. Gao, X. Yang, W. L. Dai, Y. Le, H. Li, K. Fan, *J. Catal.* **2008**, *256*, 259–267.

[181] U.S. Patent 5,155,247 W. A. Herrmann, D. M. Fritz-Meyer-Weg, M. Wagner, J. G. Kuchler, G. Weichselbaumer, R. W. Fischer, **1992**.

[182] a) R. Saladino, V. Neri, A. R. Pelliccia, R. Caminiti, C. Sadun, *J. Org. Chem.* **2002**, *67*, 1323–1332; b) R. Saladino, V. Neri, A. R. Pelliccia, E. Mincione, *Tetrahedron* **2003**, *59*, 7403–7408; c) G. Soldaini, *Synlett* **2004**, 1849–1850; d) R. Saladino, R. Bernini, V. Neri, C. Crestini, *Appl. Catal. A: General.* **2009**, *360*, 171–176.

[183] a) W. Adam, C. R. Saha-Möller, O. Weichold, *J. Org. Chem.* **2000**, *65*, 5001–5004; b) W. Adam, C. R. Saha-Möller, O. Weichold, *J. Org. Chem.* **2000**, *65*, 2897–2899.

[184] a) A.O. Bouh, J.H. Espenson, *J. Mol. Catal. A* **2003**, *200*, 43; b) M. Li, J.H. Espenson, *J. Mol. Catal. A* **2004**, *208*, 123.

[185] R. Neumann, T.J. Wang, *J. Chem. Soc. Chem. Commun.* **1997** 1915.

[186] C.D. Nunes, M. Pillinger, A.A. Valente, I.S. Gonc‚alves, J. Rocha, P. Ferreira, F.E. Kühn, *Eur. J. Inorg. Chem.* **2002** 1100.

[187] D. Veljanovski, A. Sakthivel, W.A. Herrmann, F.E. Kühn, *Adv. Synth. Catal.* **2006**, *348*, 1752.

[188] L.M.R. González, A.L.P. de Villa, C. Consuelo de Montes, G. Gelbard, *React. Funct. Polym.* **2005**, *65*, 169.

[189] I. L. V. Rosa, C. M. C. P Manso, O. A. Serra, Y. Iamamoto, *J. Mol. Catal. A: Chem.* **2000**, *160*, 199–208.

[190] L. P. B. Lôvo, F. C. Skrobot, G. C. Azzellini, Y. Iamamoto, I. L. V. Rosa, *Mod. Res. Catal.* **2013**, *2*, 47-55.

[191] L. Zhang, T. Sun, and J. Y. Ying, *J. Chem. Soc., Chem. Commun.* **1999**, 1103–1104.

[192] B.-Z. Zhan and X.-Y. Li, *J. Chem. Soc., Chem. Commun.* **1998**, 349–350.

[193] M. A. Martinez-Lorente, P. Battioni, W. Kleemiss, J.F. Bartoli, D. Mansuy, *J. Mol. Catal. A: Chem.* **1996**, *113*, 343–353.

[194] S. Tangestaninejad and V. Mirkhani, *J. Chem. Res., Synop.* **1998**, 788–789.

[195] P. Bhyrappa, J. K. Young, J. S. Moore, K. S. Suslick, *J. Am. Chem. Soc.* **1996**, *118*, 5708–5711.

[196] P. R. Cooke, J. R. L. Smith, J. Chem. Soc., Perkin Trans. 1 1913–1923 (1994).

[197] T. G. Traylor, Y. S. Byun, P. S. Traylor, P. Battioni, D. Mansuy, *J. Am. Chem. Soc.* **1991**, *113*, 7821–7823.

[198] H. Turk, W. T. Ford, J. Org. Chem. 1991, 56, 1253–1260.

[199] P. Anzenbacher Jr., V. Kril, K. Jursikova, J. Giinterovi, A. Kasal, *J. Mol. Catal. A: Chem.* **1997**, *118*, 63–68.

[200] F. Minutolo, D. Pini, and P. Salvadori, *Tetrahedron Lett.* **1996**, *37*, 3375–3378.

[201] L. Canali, E. Cowan, H. Deleuze, C. L. Gibson, D. C. Sherrington, *J. Chem. Soc., Perkin Trans. 1* **2000**, *13*, 2055–2066.

[202] B. B. De, B. B. Lohray, S. Sivaram, P. K. Dhal, *J. Polym. Sci., Part A: Polym. Chem.* **1997**, *35*, 1809–1818.

[203] M. J. Sabater, A. Corma, A. Domenech, V. Fornés, H. García, *Chem. Commun.* **1997**, 1285–1286.

[204] S. B. Ogunwumi, T. Bein, *Chem. Commun.* **1997**, 901–902.

[205] G.-J. Kim, J.-H. Shin, *Tetrahedron Lett.* **1999**, *40*, 6827–6830.

[206] D. E. De Vos, J. L. Meinershagen, T. Bein, *Angew. Chem., Int. Ed. Engl.* **1996**, *35*, 2211–2213.

[207] D. E. De Vos, S. de Wildeman, D. F. Sels, P. J. Grobet, P. A. Jacobs, *Angew. Chem., Int. Ed. Engl.* **1999**, *38*, 980–983.

[208] B. F. Sels, A. L. Villa, D. Hoegaerts, D. E. De Vos, P. A. Jacobs, *Top. Catal.* **2000**, *13*, 223–229.

[209] Y. V. S. Rao, D. E. De Vos, B. Wouters, P. J. Grobet, P. A. Jacobs, *Stud. Surf. Sci. Catal.* **1997**, *110*, 973–980.

[210] T. S. Reger, K. D. Janda, *J. Am. Chem. Soc.* **2000**, *122*, 6929–6934.

[211] C. Nozaki, C. Lugmair, A. Bell, T. D. Tilley, *J. Am. Chem. Soc.* **2002**, *124*, 13194.

[212] a) ChemSystems, *PERP Report* "Propylene Oxide" 07/08–6, Nov. **2008**, Nexant Inc; b) *Kirk-Othmer: Encyclopedia of Chemical Technology*, Richey, W. F. Chlorohydrins, 4th Edition; Wiley: New York, **1994**, 6, 140.

[213] U.S. Patent No. 6,646,139, T. Seo, J. Tsuji, (Sumitomo Corporation), **2003**.

[214] U.S. Patent No. 6,504,038, 2003, J. J. Van Der Sluis, (Shell Corporation), **2003**.

[215] U.S. Patent No. 4,891,437, E. T. Marquis, K. P. Keating, J. F. Knifton, W. A. Smith, J. R. Sanderson, J. Lustri, (Texaco Corporation), **1990**.

[216] US Patent 6,479,680, P. Bassler, W. Harder, P. Resch, N. Rieber, W. Ruppel, J. H. Teles, A. Walch, A. Wenzel, P. Zehner, **2002**.

[217] S. Bordiga, F. Bonino, A. Damin, C. Lamberti, *Phys. Chem. Chem. Phys.* **2007**, *9*, 4854–4878.

[218] C. Lamberti, S. Bordiga, A. Zecchina, G. Artioli, G. L. Marra and G. Spanó, *J. Am. Chem. Soc.* **2001**, *123*, 2204.

[219] S. Klein, S. Thorimbert, W. F. Maier, *J. Catal.* **1996**, 163, 476.

[220] a) D. C. M. Dutoit, M. Schneider and A. Baiker, *J. Catal.*, **1995**, *153*, 165; b) R. Hutter, T. Mallat, A. Baiker, *J. Catal.* **1995**, *153*, 177.

[221] E. Jorda, A. Tuel, R. Teissier and J. Kervennal, J. Catal., 1998, 175, 93.

[222] E. Gianotti, C. Bisio, L. Marchese, M. Guidotti, N. Ravasio, R. Psaro, S. Coluccia, *J. Phys. Chem. C*, **2007**, *111*, 5083.

[223] W. Adam, A. Corma, H. Garcia, O. Weichold, *J. Catal.* **2000**, *196*, 339.

[224] A. Corma, U. Diaz, V. Fornes, J. L. Jorda, M. Domine, F. Rey, *Chem. Commun.* **1999**, 779.

[225] a) O. A. Kholdeeva, T. A. Trubitsina, R. I. Maksimovskaya, A. V. Golovin, W. A. Neiwert, B. A. Kolesov, X. López, J. M. Poblet, Inorg. Chem., 2004, 43, 2284; b) P. Jiménez-Lozano, I. D. Ivanchikova, O. A. Kholdeeva, J. M. Poblet, J. J. Carbo, *Chem. Commun.* **2012**, *48*, 9266–9268.

[226] F. X. Gao, T. Yamase and H. Suzuki, *J. Mol. Catal. A: Chem.* **2002**, *180*, 97.

[227] a) A. O. Bouh, A. Hassan, S. L. Scott, *Catal. Org. Rea.* **2003**, *44*, 537; b) S. Sensarma, A. O. Bouh, S. L. Scott, *J. Mol. Catal. A Chem.* **2003**, *203*, 145.

[228] a) C. A. Bradley, M. J. McMurdo, T. Don Tilley, *J. Phys. Chem. C* **2007,** *111,* 17570-17579; b) J. Jarupatrakorn, T. Don Tilley, *J. Am. Chem. Soc.* **2002**, *124*, 8380-8388.

[229] a) P. J. Cordeiro, T. Don Tilley, *ACS Catal.* **2011**, *1*, 455-467; b) R. L. Brutchey, B. V. Mork, D. J. Sirbuly, P. Yang, T. Don Tilley, *Langmuir* **2005**, *21*, 9576-9583.

[230] R. L. Brutchey, D. A. Ruddy, L. K. Andersen, T. Don Tilley, *J. Mol. Catal. A: Chem.*, **2005**, *238*, 1–12.

[231] X. F. Li, H. X. Gao, G. J. Jin, L. Ding, L. Chen, H. Y. Yang, X. He, Q. L. Chen, *Chin. Chem. Let.* **2007**, *18*, 591-594.

[232] S. Wang, Q. Yang, Z. Wu, M. Li, J. Lua, Z. Tan, C. Li, *J. Mol. Catal. A: Chem.* **2001**, *172*, 219–225.

[233] L. Marchese, E. Gianotti, V. Dellarocca, T. Maschmeyer, F. Rey, S. Coluccia, J. M. Thomas, *Phys. Chem. Chem. Phys.* **1999**, *1*, 585-592.

[234] A. Corma, M. A. Camblor, P. Esteve, A. Martínez, J. Pérz-Pariente, *J. Catal.* **1994**, *145*, 151-158.

[235] J. H. Van der Waal, M. S. Rigutto and H. van Bekkum, *Appl. Catal. A*, **1998**, *167*, 331.

[236] M. D'Amore, S. Schwarz, *Chem. Commun.*, **1999**, 121.

[237] F. Figueras, H. Kochkar and S. Caldarelli, *Microporous Mesoporous Mater.*, 2000, **39**, 249.

[238] P. Wu, T. Tatsumi, *J. Phys. Chem. B*, 2002, **106**, 748.

[239] B. M. Choudary, V. L. K. Valli, A. D. Prasad, *J. Chem. Soc., Chem. Commun.* **1990**, 1186–1187.

[240] J. K. Karjalainen, O. E. O. Hormi, and D. C. Sherrington, *Tetrahedron: Asymmetry* **1998**, *9*, 3895–3901.

[241] a) R. Millini, E. P. Massara, G. Perego, G. Bellussi, *J. Catal.* **1992**, *137*, 497; b) G. Perego, G. Bellussi, C. Corno, M. Taramasso, F. Buonuomo, A. Esposito, *Stud. Surf. Sci. Catal.* **1987**, *28*, 129.

[242] a) S. Bordiga, S. Coluccia, C. Lamberti, L. Marchese, A. Zecchina, F. Boscherini, F. Buffa, F. Genoni, G. Leofanti, G. Petrini, G. Vlaic, *J. Phys. Chem.* **1994**, *98*, 4125; b) S. Bordiga, F. Boscherini, S. Coluccia, F. Genoni, C. Lamberti, G. Leofanti, L. Marchese, G. Petrini, G. Vlaic, A. Zecchina, *Catal. Lett.* **1994**, *26*, 195.

[243] a) A. Zecchina, M. Rivallan, G. Berlier, C. Lamberti, G. Ricchiardi, *Phys. Chem. Chem. Phys.* **2007**, *9*, 3483; b) C. Lamberti, G. Turnes Palomino, S. Bordiga, A. Zecchina, G. Spanó, C. Otero Areán, *Catal. Lett.* **1999**, *63*, 213.

[244] a) M. R. Boccuti, K. M. Rao, A. L. G. Zecchina, G. Petrini, *Stud. Surf. Sci. Catal.* **1989**, *48*, 133; b) A. Zecchina, G. Spoto, S. Bordiga, M. Padovan, G. Leofanti, *Stud. Surf. Sci. Catal.* **1991**, *65*, 671; c) M. A. Camblor, A. Corma, J. erezpariente, *J. Chem. Soc. Chem. Commun.* **1993**, 557.

[245] D. Scarano, A. Zecchina, S. Bordiga, F. Geobaldo, G. Spoto, G. Petrini, G. Leofanti, M. Padovan, G. Tozzola, *J. Chem. Soc. Faraday Trans.* **1993**, *89*, 4123.

[246] a) E. Borello, C. Lamberti, S. Bordiga, A. Zecchina, C. Otero Areán, *Appl. Phys. Lett.* **1997**, *71*, 2319

[247] C. Lamberti, *Microporous Mesoporous Mater.* **1999**, *30*, 155.

[248] R. Millini, E. P. Massara, G. Perego, G. Bellussi, *J. Catal.* **1992**, *137*, 497.

[249] S. Bordiga, Francesca Bonino, Alessandro Damin, Carlo Lamberti, *Phys. Chem. Chem. Phys.* **2007**, *9*, 4854-4878.

[250] R. Millini, G. Perego, D. Berti, W. O. Parker, A. Carati, G. Bellussi, *Microporous Mesoporous Mater.* **2000**, *35–36*, 387.

[251] C. Lamberti, G. Turnes Palomino, S. Bordiga, D. Arduino, A. Zecchina, G. Vlaic, *Jpn. J. Appl. Phys. Part 1* **1999**, *38*, 55.

[252] S. Bordiga, I. Roggero, P. Ugliengo, A. Zecchina, V. Bolis, G. Artioli, R. Buzzoni, G. L. Marra, F. Rivetti, G. Spano´ and C. Lamberti, *J. Chem. Soc., Dalton Trans.* **2000**, 3921.

[253] a) Z. Ruzic-Toros, B. Kojic-Prodic, M. Sljukic, *Inorg. Chim. Acta* **1984**, *86*, 205; b) J. R. Hagadorn, J. Arnold, *Organometallics* **1998**, *17*, 1355.

[254] F. Bonino, A. Damin, G. Ricchiardi, M. Ricci, G. Spanó, R. D'Aloisio, A. Zecchina, C. Lamberti Prestipino, S. Bordiga, *J. Phys. Chem. B* **2004**, *108*, 3573.

[255] C. Prestipino, F. Bonino, A. Usseglio Nanot, A. Damin, A. Tasso, M. G. Clerici, S. Bordiga, F. D'Acapito, A. Zecchina, C. Lamberti, *ChemPhysChem* **2004**, *5*, 1799.

[256] a) J. Wahlen, B. Moens, D. E. De Vos, P. L. Alsters, P. A. Jacobs, *Adv. Synth. Catal.* **2004**, *346*, 333 – 338; b) H. Li, Q. Lei, X. Zhang, J. Suo, *ChemCatChem* **2011**, *3*, 143 – 145; c) J. Zhuang, G. Yang, D. Ma, X. Lan, X. Liu, X. Han, X. Bao, U. Mueller, *Angew. Chem. Int. Ed.* **2004**, *43*, 6377 –6381; d) R. Palkovits, W. Schmidt, Y. Ilhan, A. Erdem-Senatalar, F. Schüth, *Microporous Mesoporous Mater.* **2009**, *117*, 228–232.

[257] S. Baek Shin, D. Chadwick, *Ind. Eng. Chem. Res.* **2010**, *49*, 8125–8134.

[258] C.-J. Liu, W. Y. Yu, S. G. Li, C. M. Che, *J. Org. Chem.* **1998**, *63*, 7364–7369.

[259] X.-Q. Yu, J.-S. Huang , W.-Y. Yu , C.-M. Che, *J. Am. Chem. Soc.* **2000**, *122*, 5337–5342.

[260] N. Scotti, N. Ravasio, F. Zaccheria, R. Psaro, C. Evangelisti, *Chem. Commun.*, **2013**, *49*, 1957.

[261] R. L. Brutchey, C. G. Lugmair, L. O. Schebaum, T. Don Tilley, *J. Catal.* **2005**, *229*, 72–81.

[262] D. A. Ruddy, T. Don Tilley, *Chem. Commun.* **2007**, 3350–3352.

[263] T. A. Nijhuis, M. Makkee, J. te A. Moulijn, B. M. Weckhuysen, *Ind. Eng. Chem. Res.* **2006**, *45*, 3447.

[264] a) M. S. Chen, D. W. Goodman, *Science* **2004**, *306* (5694), 252-255; b) T. A. Nijhuis, T. Visser, B. M. Weckhuysen, *Angew. Chem., Int. Ed.* **2005**, *44*, 1115-1118; c) C. T. Campbell, *Science* **2004**, *306* (5694), 234-235; d) S. Arrii, F. Morfin, A. J. Renouprez, J. L. Rousset, *J. Am. Chem. Soc.* **2004**, *126*, 1199-1205; e) M. L. Kimble, A. W. Castleman, R. Mitric, C. Burgel, V. Bonacic-Koutecky, *J. Am. Chem. Soc.* **2004**, *126*, 2526-2535; f) J. Guzman, B. C. Gates,. *J. Am. Chem. Soc.* **2004**, *126*, 2672-2673; g) N. Lopez, T. V. W. Janssens, B. S. Clausen, Y. Xu, M. Mavrikakis, T. Bligaard, J. K. Norskov, *J. Catal.* **2004**, *223*, 232-235; h) D. C. Meier, D. W. Goodman, *J. Am. Chem. Soc.* **2004**, *126*, 1892-1899; i) R. Zanella, C. Louis, S. Giorgio, R. Touroude, *J. Catal.* **2004**, *223*, 328-339; j) B. Schumacher, Y. Denkwitz, V. Plzak, M. Kinne, R. J. Behm,

J. Catal. **2004**, *224*, 449-462; k) M. Date, M. Okumura, S. Tsubota, M. Haruta, M. Angew. Chem., Int. Ed. **2004**, *43*, 2129-2132.

[265] T. Hayashi, K. Tanaka, M. Haruta, *J. Catal.* **1998**, *178*, 566-575.

[266] D. Gajan, K. Guillois, P. Delichère, J.-M. Basset, J.-P. Candy, V. Caps, C. Copéret, A. Lesage, L. Emsley, *J. Am. Chem. Soc.* **2009**, *131*, 14667–14669.

[267] A. Corma, H. Garcia, *Chem. Rev.* **2003**, *103*, 4307-4365

[268] J. K. F. Buijink, J. J. M. Van Vlaanderen, M. Crocker, F. G. M. Niele, *Catal. Today* **2004**, *93-95*, 199.

[269] M.C. Capel-Sanchez, G. Blanco-Brieva, J. M. Campos-Martin, M. P. de Frutos, W. Wen, J. A. Rodriguez, J. L. G. Fierro, *Langmuir* **2009**, *25*, 7148-7155;.

[270] S. Sensarma, A. O. Bouh, S. L. Scott, H. Alper, *J. Mol. Cat. A: Chem.* **2003**, *203*, 145-152.

[271] a) A. O. Bouh, G. L. Rice, S. L. Scott, *J. Am. Chem.. Soc.* **1999**, *121*, 7201-7210; b) M. G. Reichmann, A. T. Bell, *Langmuir* **1987**, *3*, 111-116; c) J. Lu, K. M. Kosuda, R. P. Van Duyne, P. C. Stair, *J. Phys. Chem. C* **2009**, *113*, 12412-12418; d) P. Serp, P. Kalck, R. Feurer, *Chem. Rev.* **2002**, *102*, 3082-3128; e) K. Schrijnemakers, P. Van Der Voort, E. F. Vansant, *Phys. Chem. Chem. Phys.* **1999**, *1*, 2569-2572; f) S. Haukka, E.-L. Lakomaa, O. Jylhä, J. Vilhunen, S. Hornytzkyj, *Langmuir* **1993**, *9*, 3497-3506; g) S. Haukka, E.-L. Lakomaa, T. Suntola, *Thin Solid Films* **1993**, *225*, 280-283.

[272] Y. Román-Leshkov, M. E. Davis, *ACS Catal.* **2011**, *1*, 1566-1580.

[273] B. A. Morrow, D. T. Molapo, in *Infrared Studies of Chemically Modified Silica* (Eds. H. E. Bergna, W. O. Roberts), CRC Press **2005**, 287-293.

[274] a) F. Rascón, R. Wischert, C. Copéret, *Chem. Sci.* **2011**, *2*, 1449-1456.

[275] a) C. Copéret, M. Chabanas, R. P. Saint-Arroman, J. M. Basset, *Angew. Chem. Int. Ed.* **2003**, *42*, 156-181.

[276] a) M. Tada, Y. Iwasawa, *Coordin. Chem. Rev.* **2007**, *251*, 2702-2716, b) M. Tada, K. Motokura, Y. Iwasawa, *Top. Catal.* **2008**, *48*, 32-40, c) S. L. Wegener, T. J. Marks, P. C. Stair, *Acc. Chem. Res.* **2012**, *45*, 206-214.

[277] a) L. T. Zhuravlev, *Colloids and Surfaces A* **2000**, *173*, 1-38; b) M. Armandi, V. Bolis, B. Bonelli, C. Otero Arean, P. Ugliengo, E. Garrone, *J. Phys. Chem. C* **2011**, *115*, 23344-23353.

[278] a) C. W. Yoon, K. F. Hirsekorn, M. L. Neidig, X. Yang, T. D. Tilley, *ACS Catal.* **2011**, *1*, 1665-1678, b) X. Solans-Monfort, J.-S. Filhol, C. Copéret, O. Eisenstein, *New J. Chem.* **2006**, *30*, 842-850.

[279] S. Haukka, E. L. Lakomaa, A. Root, *J. Phys. Chem.* **1993**, *97*, 5085.

[280] a) S. D. Fleischman, S. L. Scott, *J. Am. Chem. Soc.* **2011**, *133*, 4847–4855; b) R. N. Kerber, A. Kermagoret, E. Callens, P. Florian, D. Massiot, A. Lesage, C. Copéret, F. Delbecq, X. Rozanska, P. Sautet, *J. Am. Chem. Soc.* **2012**, *134*, 67767-6775.

[281] a) K. T. Li, I. C. Chen, *Ind. Eng. Chem. Res.* **2002**, *41*, 4028; b) K. T. Li, C. C. Lin, *Catal. Today* **2004**, *97*, 257; c) K. T. Li, P. H. Lin, S. W. Lin, *Appl. Catal. A: Gen.* **2006**, *301*, 59–65.

[282] a) E. I. Ross-Medgaarden, E. I. Wachs, I. *J. Phys. Chem. C* **2007**, *111*, 15089-15099, b) X. Gao, I. E. Wachs, *Catal. Today* **1999**, *51*, 233-254, c) X. Gao, S: R. Bare, J. L. G. Fierro, M. A. Banares, I. E. Wachs, *J. Phys. Chem. B* **1998**, *102*, 5653-5666.

[283] L. A. O'Dell, R. W. Schurko, *Chem. Phys. Lett.* **2008**, *464*, 97–102.

[284] H. Hamaed, J. M. Pawlowski, B. F. T. Cooper, R. Fu, S. H. Eichhorn, R. W. Schurko, *J. Am. Chem. Soc.* **2008**, *130*, 11056-11065.

[285] S. Adiga, D. Aebi, D. L. Bryce, *Can. J. Chem.* **2007**, *85*, 496-505

[286] a) B. A. Morrow, I. A. Cody, *J. Phys. Chem.* **1976**, *80*, 1998-2004; b) B. C. Bunker, D. M. Haaland, Michalske, W. L. Smith, *Surface Sc.* **1989**, *222*, 95-118.

[287] a) J. M. Fraile, J. García, J. A. Mayoral, M. Grazia Proietti, M. C. Sánchez *J. Phys. Chem.* **1996**, *100*, 19484-19488, b) E. A. Stern, Am. Phys. Soc. **1993**, *48*, 9825.

[288] a) C. A. Demmelmaier, R. E. White, J. A. van Bokhoven, S. L. Scott, *J. Phys. Chem. C.* **2008**, *112*, 6439-6449; b) C. A. Demmelmaier, R. E. White, J. A. van Bokhoven, S. L. Scott, *J. Catal.* **2009**, *262*, 44-56.

[289] J. Guo, L. Mao, J. Zhang, C. Feng, *Appl. Surf. Sc.* **2010**, *256*, 2132–2137.

[290] I. Bello, W. H. Chang, W. M. Lau, *J. Appl. Phys.* **1994**, *75*, 3092-3097.

[291] B. P. C. Herreijgers, R. F. Parton, B. M. Weckhuysen, *ACS Catal.* **2011**, *1*, 1183-1192.

[292] Gaussian 09, Revision A.02, Frisch, M. J.; Trucks, G. W.; Schlegel, H. B.; Scuseria, G. E.; Robb, M. A.; Cheeseman, J. R.; Scalmani, G.; Barone, V.; Mennucci, B.; Petersson, G. A.; Nakatsuji, H.; Caricato, M.; Li, X.; Hratchian, H. P.; Izmaylov, A. F.; Bloino, J.; Zheng, G.; Sonnenberg, J. L.; Hada, M.; Ehara, M.; Toyota, K.; Fukuda, R.; Hasegawa, J.; Ishida, M.; Nakajima, T.; Honda, Y.; Kitao, O.; Nakai, H.; Vreven, T.; Montgomery, J. A. Jr.; Peralta, J. E.; Ogliaro, F.; Bearpark, M.; Heyd, J. J.; Brothers, E.; Kudin, K. N.; Staroverov, V. N.; Kobayashi, R.; Normand, J.; Raghavachari, K.; Rendell, A.; Burant, J. C.; Iyengar, S. S.; Tomasi, J.; Cossi, M.; Rega, N.; Millam, J. M.; Klene, M.; Knox, J. E.; Cross, J. B.; Bakken, V.; Adamo, C.; Jaramillo, J.; Gomperts, R.; Stratmann, R. E.; Yazyev, O.; Austin, A. J.; Cammi, R.; Pomelli, C.; Ochterski, J. W.; Martin, R. L.; Morokuma, K.; Zakrzewski, V. G.; Voth, G. A.; Salvador, P.; Dannenberg, J. J.; Dapprich, S.; Daniels, A. D.; Farkas, O.; Foresman, J. B.; Ortiz, J. V.; Cioslowski, J.; Fox, D. J. Gaussian, Inc., Wallingford CT, **2009**.

[293] a) A. D. Becke, *J. Chem. Phys.* **1992**, *96*, 2115; A. D. Becke, *J. Chem. Phys.* **1992**, *97*, 9173. A. D. Becke, *J. Chem. Phys.* **1993**, *98*, 5648; b) C. Lee, W. Yang, R. G. Parr, *Phys. Rev. B* **1988**, *37*, 785.

[294] A. W. Ehlers, M. Boehme, S. Dapprich, A. Gobbi, A. Hoellwarth, V. Jonas, K. F. Koehler, R. Stegmann, A. Veldkamp, G. Frenking, *Chem. Phys. Lett.* **1993**, *208*, 111–114.

[295] A. P. Scott, L. Radom, *J. Phys. Chem.* **1996**, *100*, 16502.

[296] J. Tomasi, B. Mennucci, R. Cammi, *Chem. Rev.* **2005**, *105*, 2999.

[297] M. Abon, J.-C. Volta, *Appl. Catal. A: Gen.*, **1997**, *157*, 173.

[298] G. Calleja, J. Aguado, A. Carrero, J. Moreno, *Appl. Catal. A: Gen.* **2007**, *316*, 22

[299] M.P. McDaniel, G. Ertl, H. Knözinger, F. Schüth, J. Weitkamp, *Handbook of Heterogeneous Catalysis*, Wiley–VCH, Weinheim, 2008, 3733.

[300] P.M. Morse, *Chem. Eng. News* **1999**, *77*, 11.

[301] J. A. N. Ajjou, G. L. Rice, S. L. Scott *J. Am. Chem. Soc.* **1998**, *120*, 13436.

[302] B.M. Weckhuysen, D.E. Keller, *Catal. Today* **2003**, *78*, 25.

[303] I.E. Wachs, *Catal. Today* **2005**, *100*, 79.

[304] M.A. Bañares, *Catal. Today* **1999**, *51*, 319.

[305] T. Blasco, J.M.L. Nieto, *Appl. Catal., A* **1997**, *157*, 117.

[306] S. L. Scott, J. Amor Nait Ajjou, *Chem. Eng. Sci.* **2001**, *56*, 4155.

[307] M. A. Vuurmant, I. E. Wachs, *J. Phys. Chem.* **1992**, *96*, 5008.

[308] B. Morosin, A. Narath, *J. Chem. Phys.* **1964**, *40*, 1958.

[309] a) J. A. Barth, *Z. anorg. Allg. Chem.* **1975**, *414*, 109; b) N. Goldberg, B. S. Ault, *J. Mol. Struct.* **2005**, *749*, 84.

[310] W. E. Hobbs, *J. Chem. Phys.* **1958**, *28*, 1220.

[311] F. Tabak, A. Lascialfari, A. Rigamonti, *J. Phys.: Condens. Matter* **1993**, *5*, B31; H. Zeltmann, L.O. Morgan, *Inorg. Chem.* **1971**, *10*, 2739.

[312] M. D. Zidan, A. W. Allaf, *Spectrochim. Acta, Part A*, **2000**, *56*, 2693.

[313] J. A. Barth, *Z. anorg. Allg. Chem.* **1975**, *414*, 109.

[314] B. Horvath, J. Geyer, H.L. Krauss, *Z. anorg. Allg. Chem.*, **1976**, *426*, 141.

[315] W.C. Vining, J. Strunk, A.T. Bell, *J. Catal* **2011**, *281*, 222.

[316] E. W. Deguns, Z. Taha, G. D. Meitzner, S, L. Scott, *J. Phys. Chem. B* **2005**, *109*, 5005.

[317] K.P. Bryliakov, N.N. Karpyshev, S.A. Fominsky, A.G. Tolstikov, E.P. Talsi, J. *Mol. Catal. A: Chem.* **2001**, *171*, 73.

[318] Javier Ruiz, M. Vivanco, C. Floriani, A. Chiesi-Villa, C. Rizzoli, *Organometallics* **1993**, *12*, 1811.

[319] M. Baron, H. Abbott, O. Bondarchuk, D. Stacchiola, A. Uhl, S. Shaikhutdinov, H.-J. Freund, Cristina Popa, Maria V. Ganduglia- Pirovano, J. Sauer *Angew. Chem. Int. Ed.* **2009**, *48*, 8006.

[320] G. L. Rice, S, L. Scott, *Langmuir*, **1997**, *13*, 1545.

[321] H. Yun, J. Li, H.-B. Chen, C.-J. Lin, *Electrochimica Acta* **2007**, *52*, 6679.

[322] R. Rulkens, J. L. Male, K. W. Terry, B. Olthof, A. Khodakov, A. T. Bell, E. Iglesia, T. Don Tilley, *Chem. Mater.* **1999**, *11*, 2966.

[323] a) P. Sabatier, J.-B. Senderens, C. R. Hebd. Seances *Acad. Sci.* **1897**, *124*, 616; b) P. Sabatier, J.-B. Senderens, C. R. Hebd. Seances Acad. Sci. **1897**, *124*, 1358; c) *Handbook of Heterogeneous Catalysis*, Vol. 2 (Eds: G. Ertl, H.Knözinger, J. Weitkamp), **1997**, Wiley-VCH, Weinheim; d) C. Copéret, J. Thivolle-Cazat, J.-M. Basset, *Fine Chemicals through Heterogeneous Catalysis* (Eds.: R. A. Sheldon, H. van Bekkum), **2001**, Wiley-VCH, Weinheim, p. 553.

[324] a) M. H. Ab Rahim, M. M. Forde, R. L. Jenkins, C. Hammond, Q. He, N. Dimitratos, J. A. Lopez-Sanchez, A. F. Carley, S. H. Taylor, D. J. Willock, D. M. Murphy, Christopher J. Kiely, G. J. Hutchings; *Angew. Chem. Int.l Ed.*, **2013**, *52*, 1280; b) C. Hammond, M. M. Forde, M. H. Ab Rahim, A. Thetford, Q. He, R. L. Jenkins, N. Dimitratos, J. A. Lopez-Sanchez, N. F. Dummer, D. M. Murphy, A. F. Carley, S. H. Taylor, D. J. Willock, E. E. Stangland, J. Kang, H. Hagen, C. J. Kiely, G. J. Hutchings, *Angew. Chem. Int.l Ed.*, **2012**, *51*, 5129.

[325] G. L. Rice, S. L. Scott, *Chem. Mater.*, **1998**, 10, 620

[326] G. Ricchiardi, A. Damin, S. Bordiga, C. Lamberti, G. Spanò, F. Rivetti, Adriano Zecchina, *J. Am. Chem. Soc.* **2001**, *123*, 11409-11419.

[327] M. Beaudoin, S. L. Scott, *Organometallics* **2001**, *20*, 237-239.

[328] P. F. Henry, M. T. Weller, C. C. Wilson, *J. Phys. Chem. B* **2001**, *105*, 7452.

[329] P. Ratnasamy, D. Srinivas, H. Knözinger, *Adv. Catal.* **2004**, *48*, 1.

[330] L. Lefort, M. Chabanas, O. Maury, D. Meunier, C. Copéret, J. Thivolle-Cazat, J.-M. Basset, *J. Organomet. Chem.* **2000**, *96*, 593-594.

[331] F. Lefebvre, J.-M. Basset, *J. Mol.Catal. A: Chem.* **1999**, *146*, 3-12.

[332] M. Chabanas, V. Vidal, C. Copéret, J. Thivolle-Cazat, J.-M. Basset, *Angew. Chem. Int. Ed.* **2000**, *39*, 1962.

[334] F. Blanc, C. Copéret, A. Lesage, L. Emsley, *Chem. Soc. Rev.* **2008**, *37*, 518.

[335] R.L. Brutchey, B. V. Mork, D. J. Sirbuly, P. Yang, T. Don Tilley, *J. Mol. Catal. A: Chem.* **2005**, *238*, 1–12.

[336] a) Wendelin J. Stark, Sotiris E. Pratsinis, and Alfons Baiker, *J. Catal.* **2001**, *203*, 516–524; b) J. Yao, W. Zhan, X. Liu, Y. Guo, Y. Wang, Y. Guo, G. Lu, *Microporous Mesoporous Mater.* **2012**, *148*, 131–136; c) Wu, P.; Tatsumi, T.; Komatsu, T.; Yashima, T. *Chem. Mater.* **2002**, *14*, 1657.

[337] G. Maier, H. P. Reisenauer, K. Schöttler, U. Wessolek-Kraus, *J. Organomet. Chem.* **1989**, *366*, 25-38.

[338] R. T. Carlin, R. A. Osteryoung, J. S. Wilkes, J. Rovang, *Inorganic Chemistry* **1990**, *29*, 3003–3009.

[339] H.-H. Schmidtke, U. Voets, *Inorg. Chem.* **1981**, *20*, 2766–2771.

[340] C. Li, G. Xiong, J. Liu, P. Ying, Q. Xin, Z. Feng, *J. Phys. Chem. B* **2001**, *105*, 2993-2997.

[341] S. Bordiga, F. Geobaldo, C. Lamberti, A. Zecchina, F. Boscherini, F. Genoni, G. Leofanti, G. Petrini, M. Padovan, S. Geremia, G. Vlaic, Nucl. Instrum. *Methods Phys. Res., Sect. B* **1995**, *97*, 23.

[342] G. B. van der Voet, T. I. Todorov, J. A. Centeno, W. Jonas, J. Ives, F. G. Mullick, *Mil. Med.* **2007**, *172*, 1002.

[343] M. Guidotti, C. Pirovano, N. Ravasio, B. Lazaro, J. M. Fraile, J. A. Mayoral, B. Coq, A. Galarneau, *Green Chem.* **2009**, *11*, 1421.

[344] G. F. Thiele, E. Roland, *J. Mol. Catal. A: Chem.* **1997**, *117*, 351–356.

[345] M. Stöckmann, F. Konietzn, J. U. Notheis, J. Voss, W. Keune, W. F. Maier, *Applied Catalysis A: General* **2001**, *208*, 343–358.

[346] a) L. Y. Chen, G. K. Chuah, S. Jaenicke, *J. Mol. Catal. A: Chem.* **1998**, *132*, 281–292; b) S. Park, K. M. Cho, M. H. Youn, J. G. Seo, S. H. Baeck, T. J. Kim, Y. M. Chung, S. H. Oh, I. K. Song, *Catal. Lett.* **2008**, *122*, 349–353; c) S. Park, K. M. Cho, M. H. Youn, J. G. Seo, S. H. Baeck, T. J. Kim, Y. M. Chung, S. H. Oh, I. K. Song, *Catal. Commun.* **2008**, *9*, 2485–2488.

10 List of Abbreviations and Acronyms

B3LYP	Becke-Lee-Yang-Parr hybrid functional
CBS	complete basis set
DFT	density functional theory
EPR	electron paramagnetic resonance
GC	gas chromatography
FID	flame ionization detector
IR	infrared
MS	mass spectrometry
TS	transition state
UV	ultraviolet
VIS	visible light
ν	chain length (greek nu)
η	hapticity of a ligand (greek eta)
μ	bridging properties of ligands (greek mu)
acac	Pentan-2,4-dion
PTC	phase transfer catalyst
AIBN	2,2′-Azobis(2-methylpropionitril)
MTO	methyltrioxorhenium
e.e.	enantiomeric excess
OXONE	potassium peroxomonosulfate
TPP	tetraphenylporphyrin
CVP	Chemical Vapor Deposition
*m*CPBA	*meta*-Chloroperoxybenzoic acid
CHP	cumene hydroperoxide process
PO/SM	Shell's Propylene oxide/styrene monomer process
PO/TBA	Halcon's propylene oxide/*tert*-butyl alcohol process
HPPO	BASF/Dow's hydrogen peroxide/propene oxide process
TS-1	Titaniumsilicate-1

TEOS	Tetraethoxysilane
TPD	Temperature programmed desorption
HFIP	1,1,1,3,3,3-hexafluoro-2-propan

i want morebooks!

Buy your books fast and straightforward online - at one of world's fastest growing online book stores! Environmentally sound due to Print-on-Demand technologies.

Buy your books online at
www.get-morebooks.com

Kaufen Sie Ihre Bücher schnell und unkompliziert online – auf einer der am schnellsten wachsenden Buchhandelsplattformen weltweit! Dank Print-On-Demand umwelt- und ressourcenschonend produziert.

Bücher schneller online kaufen
www.morebooks.de

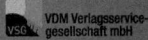 VDM Verlagsservicegesellschaft mbH
Heinrich-Böcking-Str. 6-8
D - 66121 Saarbrücken

Telefon: +49 681 3720 174
Telefax: +49 681 3720 1749

info@vdm-vsg.de
www.vdm-vsg.de

Printed by Books on Demand GmbH, Norderstedt / Germany